Careers in *Environmental Conservation*

Careers in Environmental Conservation

SIXTH EDITION

Revised by **ROBERT LAMB**

First published in 1982
Second edition 1986
Third edition 1988
Fourth edition 1989
All entitled Careers in Conservation by John McCormick
Fifth edition 1992, reprinted with revisions 1993
Sixth edition 1995, revised by Robert Lamb

Kogan Page Limited
120 Pentonville Road
London N1 9JN

British Library Cataloguing in Publication Data

A CIP record for this book is available from the British Library.

ISBN 0-7494-1567-3

Typeset by DP Photosetting, Aylesbury, Bucks
Printed and bound in Great Britain by
Biddles Limited, Guildford and Kings Lynn

Contents

Introduction

Chapter 1
What is Environmental Conservation?

In the past 100 years, the planet's human population has tripled, the global economy has grown 20 times over, world industry has expanded fiftyfold, and the burning of fossil fuels like coal and oil has increased by a factor of 30. Skyrocketing industrial growth has also meant a rush towards urbanisation, or the crowding of most of the world's people into towns or cities.

Fearsome problems and dilemmas have arisen from these changes. Activities undertaken in the name of industrial development have exhausted the very resources on which development is based, such as the world's finite metal and fossil fuel reserves, or renewable resources like forests and fisheries.

Unsustainable resource use has also depleted naturally occurring species and habitats, key strongholds of environmental stability and biodiversity. Incalculable amounts of genetic information - the stuff of evolution and of future options on human progress - have been lost in the process.

Runaway resource consumption has also caused more general blight. Chemicals and wastes from factories, farms, municipal sewage systems and power stations have polluted the air, land, seas and inland waters, notably in the heavily industrialised global 'North', though they are now obvious elsewhere. In the end, such ills are incompatible with the wealth of nations.

Local Concerns: World Problems

In the developing countries of the global 'South' the heaviest environmental costs have so far been associated with the ruin of natural 'resource bases', not least through forest loss, soil erosion and the spread of deserts. The North can generally afford to buy its way out of such crises but it has not escaped 'acid rain' damage to major forest and freshwater systems, caused by exhausts from industry and transport.

Modern environmental consciousness is a direct reaction to issues like these and their recent consequences. It took wing when the mass media began multiplying the pro-environment message in the 1970s, but had hatched long before.

The history of environmental conservation in Britain dates back at least to 1866, when the Commons Protection Society (now the Open Spaces Society) was set up to rally public dissent against the latest in a series of Enclosures Acts that had put more than six and a half million acres of communally owned fields and forests out of bounds to the citizen.

In the late 1880s a group of women in Altrincham launched a protest campaign against the killing of wild birds for the hat trade. It led to the creation of the Royal Society for the Protection of Birds, the world's foremost bird conservation body. The National Trust was born around the same time, out of public concern over railway cuttings in the Lake District.

Similarly, the birth of environmental consciousness in North America led to the creation of the world's first National Parks at Yellowstone and Yosemite in the late 1890s, after artists and writers like Church and Muir warned of the loss of America's wilderness soul, foreshadowed by the near-extinction of the plains buffalo. Informed public opinion did the rest.

In the 1920s, photographs by Ansell Adams of the Sierra Nevada mountains, multiplied by the new medium of the photo-illustrated journal, awakened even hardened city-dwellers to the spectacular natural heritage in which they held a stake. Soon after, the first (and still America's largest) pro-environment citizens' group, the Sierra Club, was born.

Much later, in the early 1960s, a marine biologist, Rachel Carson, published *The Silent Spring*, an eloquent complaint about the destruction of songbirds and other wildlife as a result of overkill use of the insecticide DDT. The outcry the book caused in and outside the USA later led to an almost worldwide ban on DDT use and a global questioning of the right by which industry and commerce evaded blame for pollution.

Plainly, then, environmental conservation originated as a field largely defined by *issues* and by the *attitudes* ordinary people adopt towards them, quite often expressed in emotive or dissident terms or through the creative vision of artists. Though hard-headed technologists might sometimes wish things otherwise, this 'wild card' will always be part of the game.

The purely technical pursuit of higher environmental standards cannot easily be separated from *how people feel* and *what they believe* about their personal kinship with nature and the wild, the quality of their immediate surroundings and the fate of their distinctive natural and cultural heritage. Professional achievers in environmental conservation should be wary of denying or ignoring such feelings and beliefs: they almost certainly owe their job to them!

Science Roots

Even so, modern environmental conservation is also at its roots a *science-based and practical* pursuit. It has used a magpie, interdisciplinary mix of sciences to refine its mission, ranging from classical natural history to applied ecology, from economics to computer or space technology.

This combination of science-based credibility and popular appeal is unique. Glamorised by the eye-catching 'media theatre' of green pressure-groups, it has shot environmental conservation into the headlines and to the head of the world political and economic agenda in recent decades.

A prototype among science-based conservation bodies was the International Union for the Protection of Nature (IUPN), now the World Conservation Union, formed in France in 1948 by a group of mainly European and American naturalists, writers, philanthropists and zoologists.

The name was soon changed to the International Union for Conservation of Nature and Natural Resources (IUCN), despite complaints from French-speaking members that to them 'conservation' meant only making jam or bottling plums!

The change was made to reflect the idea that the 'wise use' of nature and natural resources - such as the farming of wild game reared for the purpose, or the harvesting of forests on a 'sustained yield' rotation basis - could be compatible with rescuing species or safeguarding wilderness. None the less, the Union's agenda still rested mainly on the protection of natural areas, the survival of rare and endangered animals and the formulation of new legal instruments to safeguard the wild.

In the late 1950s and the 1960s, this range of concerns was again enlarged, first to reflect new knowledge about pollution and other effects of industrial and agricultural development. This shift linked conservation to a much wider technical and academic knowledge base. Hybrid terms like 'environmental conservation' came into use to redefine the field accordingly. These, too, have more recently been superseded by the larger concept of sustainable development (see page 14), which embraces the view that environmental and conservation thinking are of crucial importance, but primarily for setting new parameters to what makes the social and economic development of the world's poorest people and communities fairer, easier to achieve and less difficult to hold on to for generations.

Industry Bandwaggon

As for the private economic sector, laws and regulations started coming into force from around the 1950s on, such as Clean Air Acts in the USA and the UK, that began to impose strict obligations on industrialists, traders, developers and investors to clean up their act along with everybody else.

The Organization of Economic Cooperation and Development (OECD) agreed in 1967 on a 'Polluter Pays Principle', holding industries or businesses that caused pollution incidents or environmental public health hazards, responsible for repaying environmental protection agencies the full cost of setting matters right. Gradually this principle entered legislation.

By the mid-1970s, the first 15 or 20 of what now adds up to a wealth of more than 200 environmental directives or regulations were being approved by the governing bodies of the European Community (now the European Union), in response to a call issued at the 1972 Stockholm Conference on the Human Environment, precursor of the 1992 Earth Summit (see page 13, A Duty of Care).

The rising cost of conforming to environmental regulations, buying waste disposal licences or paying fines for non-compliance, began to turn business enterprises and industrial corporations on to focusing their research and development efforts on new product and process designs that avoided polluting by-products or waste products altogether. Some have gone further, making positive selling-points and (in most cases well-deserved) profits out of 'environmentally friendly' processes or recycled components or ingredients.

Public Service Meets Public Opinion

On the official side, government agencies and local authority departments responded in their own way to rising public clamour for better standards of environmental care, at first by changing labels on departmental doors. Most Public Health or Public Works departments around the UK became Environmental Services departments, for example. The work they housed seemed little different from their former tasks and the personnel looked familiar, too: the same orthodox architects, transport engineers, town and country planners and health inspectors.

In time, however, a new breed of specialised 'green collar worker' began to enter the public service ranks. Now most local authorities and statutory bodies boast a firm core of environmental expertise based on qualifications, training and experience specially geared or adapted to the job in hand.

This process began partly as a result of top-down legislative pressures but also in response to a youthful protest movement that emerged to prominence in the late 1960s and early 1970s. It linked life-threatening aspects of pollution and species loss to a (then) more typical anti-war and anti-nuclear lobby.

A new beacon was lit when popular advocacy and campaigning groups like Friends of the Earth, WWF and Greenpeace latched on to the huge potential social and political leverage at the command of the mass media, which soon became their willing ally. During the 1980s an average of around 2000 new environmental groups were established worldwide each year.

This upsurge of popular environmentalism was not confined to the media-saturated North. From regarding environmental conservation as a plot against their development, developing country leaders now came to see the point of learning from the North's wrong turns and seeking prosperity along more sustainable lines as a matter of paramount self-interest. In 1992, these converging trends met at Rio (pages 13 and 14).

What provided the final impetus towards this meeting of minds was a series of revelations about the largest and most important of the global 'commons', the atmosphere. Pollution by chlorine-based industrial chemicals was depleting the stratospheric ozone layer which protects the Earth from the harmful effects of the Sun's ultra-violet radiation.

More disturbing still was evidence that surplus carbon dioxide and other 'greenhouse gas' emissions caused by fossil fuel burning in industry, households and transport were forcing up air temperatures. Unless checked or reversed, this trend could cause potentially catastrophic climate changes with notably menacing implications for agriculture and mean sea levels.

No longer could anyone sit back and let the combined processes of agricultural development, industrialisation and urbanisation that had given rise to these menaces, proceed as before, without heed of their effect on the safety and livelihoods of tomorrow's generations. A new era in environmental conservation had arrived, though its impact was dulled by the world economic recession that arrived with it.

Strategic Alliances

This brief overview has shown how environmental conservation has come to life and annexed an ever-widening range of public, professional or academic concerns over the past century or so, concerns that are not necessarily compatible with one another. In fact, the only thing they can be guaranteed to have in common is that they all relate in the end to the same nature.

Yet for this same good reason, these concerns increasingly serve to seal strategic alliances across professional sectors that once seemed to have little or nothing to say to one another. This new willingness to join hands across old boundaries will be a distinctive feature of the way careers in environmental conservation develop over the next five years.

The last edition of this book was safely able to say that most career openings in environmental conservation were, in roughly ascending order of payroll potential, in the voluntary, the government or the professional sector.

In the past few years, that order has been virtually reversed but the classification has changed anyway, owing to shifts in circumstance and usage. It is now, for instance, common practice to refer to an independent rather than a voluntary sector and to include in it not only private voluntary organisations (PVOs) and non-governmental organisations (NGOs) but also environmental interests in business, commerce and industry. It also encompasses chartered professional bodies active in environmental planning or management and the many independent research institutes and consultancy partnerships that offer contractual services in a range of sectors.

The official sector incorporates government agencies, nationalised industry, statutory bodies and quasi-non-governmental organisations ('quangos') or institutions at all levels, from international bodies like the UN and regional forums like the EU, through to national and local government levels. Legal professions and institutions also fall mostly in this slot.

A significant recent departure from it in Britain is that of the privatised power and water utilities. It is quite possible that remaining statutory or state-owned bodies like the national parks authorities and the Forestry Commission may soon follow suit. Though these changes are in progress, this book does not assume they are a *fait accompli*. Its main focal area remains the UK, though helpful advice is offered to readers interested in careers in the EU as a whole or in the wider international environment and development arena.

The Structure of This Book

The section that follows surveys a selection of key issues likely to be 'hot spots' of growing environmental concern in the next five years. It leads on to chapters describing some of the main types of work activity included within the environmental conservation field as it now stands and the key qualifications and talents that would-be recruits are likely to need if they seek careers that relate to these activities. The framework around these chapters is that of two main sections dealing respectively with the independent and the official sector, and featuring leading organisations whose operations typify particular fields or ways of working.

The range of examples is not exhaustive: the Environment Council lists more than 750 separate organisations active within the UK, in its latest *Who's Who in the Environment*! Details of this and other reference sources can be found in the 'Useful Reading' section of the present work.

Other sections itemise environmental 'Courses and Qualifications' in the UK and offer a 'Useful Addresses' list. Once again, these lists are not exhaustive but reference is made in the Useful Books section to comprehensive directories and yearbooks that deal with education and training opportunities in detail. Though every effort is made to verify information included in this book, the publisher would be grateful to hear from anyone who spots omissions or mistakes.

How Many Jobs?

How many jobs are there in environmental conservation today? According to the Institution of Environmental Sciences in its *Environmental Careers Handbook*, there are at least 300,000 potential openings in the UK alone, if volunteer positions and jobs in the building trades are left out of the calculation, though many of those that remain are for manual labourers.

A feature of the rough breakdown the IES has made of other jobs by sector is the relatively fast growth of the consultancy, investment, industry and education sectors, contrasting with a near standstill (by now a definite contraction) in the private voluntary sector. The latter trend reflects recession-hit membership incomes and the end of Youth Opportunity Programmes and other job creation schemes, on which voluntary organisations relied heavily in the 1980s.

Basic levels of pay in the UK vary surprisingly little from sector to sector. In most typical non-governmental organisations of any size, professional salaries will usually range between £12,000 and £35,000 a year, in smaller organisations often less. According to the results of a 1992 survey conducted in industry by the newsletter *Environment Business*, basic salaries range from around £14,000 a year for specialists working as environmental managers on industrial sites, up to an *average* £40,000 (maximum £60,000) for Group Directors who administer environmental policy at large.

Engineering specialists earned substantially more than science specialists, while experts operating in consultancies did better than those with fixed posts in the industry hierarchy. On the industry and consultancy fronts and on the regulatory side of the official sector, recession has done less damage to employment prospects than it has elsewhere, probably because the backlog of environmental regulation and legislation demanding enforcement has kept these sectors on the move.

Footnote

One preliminary question that anyone seeking involvement in environmental conservation out of a personal sense of concern and commitment should ask themselves is: 'Do I really need a career in environmental conservation to fulfil this mission?'

We all should have realised by now that environment is an all-inclusive concept, that people who enter work of any kind able and willing to urge higher environmental standards in their chosen occupation, can do just as much as any career-specific environmentalist - if not more - to promote sustainable development, for they have the direct power to make these standards part of the enduring routine of life and livelihood.

A Duty of Care

The past 15 years have seen a radical shift in public attitudes to environmental issues and (it follows) in the role of professionals who respond to those issues and attitudes. The most significant changes are also the most recent: they are embedded in agreements settled at the UN Conference on Environment and Development - UNCED or the Earth Summit for short. The biggest international conference ever convened, UNCED took place in Rio di Janeiro, Brazil in June 1992.

UNCED's point was partly symbolic. National delegations and leaders representing all UN Member States set a global seal of approval on a view of the living world that till then had still been dismissed by many as a minority or fringe outlook.

In this view, looking after nature is not the special concern of a few narrowly focused professions or groups active in tidily separated subject-areas but is *everyone's* business, regardless of vocation, lifestyle or other distinction. Whatever our field of action at the workplace or in private, we constantly owe *a duty of care* to the environment, for most of the things we do from day to day are liable, directly or indirectly, to affect its condition for better or worse.

In practical terms, it means that efforts by specialist lawyers, planners, campaigners, technicians and others to safeguard or boost the quality of the human environment or to conserve wildlife and natural habitats, should not run counter to everyday goals pursued by ordinary people in all walks of life, or they will not last long enough to count.

Though the Earth Summit paid special heed to obstacles to environmental care that confront developing countries, *all* countries depend for their well-being on developing or acquiring a growing supply of extracted or harvested natural materials to support a human population whose numbers, expectations, or both, are constantly on the rise. They must do this without depleting resources beyond hope of recovery or disabling 'life support systems' such as clean air and water, on which all biological and other production systems depend.

Independent conservation organisations and green pressure-groups face a stern challenge in view of the shift of emphasis signalled at UNCED: to switch from a role principally dedicated to 'blowing the whistle' on environmental and conservation *problems* to a new role as practical *problem-solvers* with a special brief for grassroots participation. Professional environmental specialists in the official and economic sectors, face a converse test: coming to terms with the human factor in the management of nature and resources.

Sustainable Development

One of the Earth Summit's most significant outcomes was *Agenda 21* - an informal framework of agreed priorities for rethinking development along sustainable lines at global, national and local levels in the 21st century. It recognises that most environmental problems pay no heed to political boundaries so must be tackled strategically, on a global or regional scale, as well as by the nations or neighbourhoods most closely affected by them. Above all, it establishes that environmental concerns should be regarded as synonymous with concern for human individuals and communities.

Most people do not need to be told that their livelihoods, health standards and peace of mind come under threat whenever harm is done to their immediate environment, staple natural resources or 'common property resources' like wilderness and wildlife. But people already fighting losing battles against hunger, lack of healthcare and other life-threatening ills, hardly have much choice in the matter.

Conversely, if environmental care progresses in harness with a growth in livelihoods, an improving quality of life and a fairer social deal for ordinary people, as well as catering to market demands and the wealth of nations, its benefits should be reinforced and guaranteed by sustained economic and human benefits at every step of the way, almost without need of further safeguards or interventions.

This down-to-earth *sustainable development* argument is now beginning to supplant some more technical or more doctrinaire concepts of environmental conservation that have prevailed in the past, as a dominant motivating force behind contemporary efforts to square human progress with environmental care.

Would-be professionals in environmental conservation should have a sound working knowledge of the Earth Summit Agreements and the basic principles of sustainable development, for they may determine the future pattern of professional development in fields specific to environmental care or management in the years to come. For instance, local authorities and other environmental 'gate-holders' in the UK and EU are now responding to a call issued at Rio, to develop their own versions of *Agenda 21* tuned to local conditions and realities.

Sustainable development is also the guiding principle behind the European Union's current *Vth Action Programme for the Environment*. It plots an agreed course of development for environmental care provisions, legislation and regulation throughout the Union, as well as defining priority areas for statutory grant support under relevant financial instruments. In the past few years, most independent pro-environment organisations have been reorienting their goals in terms of sustainable development, to take advantage of this synergy.

Chapter 2

Approaching the Issues

To come to terms with environmental conservation as a professional concern, it is vital to have an overview of today's (and tomorrow's) most urgent issues. Whatever topic happens to preoccupy the conservation professional in an immediate way, he or she can feel sure that it will be interrelated with a host of others. For it is an axiom of ecological science and of environmentalism as a whole that no living system exists in an isolated state: pressure exerted at a particular point in the 'web of life' is almost bound to affect the rest of it too.

Another good reason to know one's way around the issues is that, as we have seen, progress in environmental conservation was very largely driven by issues from the beginning and there is every likelihood that the same will happen in the future. It makes sense to get a step 'ahead of the game' and be in a position to predict issues which will head tomorrow's agenda.

Nowadays, there are more clues available to inform this guessing-game than ever before. In documents like *Agenda 21*, the EU *Vth Action Programme for the Environment* or the *World Conservation Strategy* and its updates, certain officially agreed priority issues are set out in the clearest possible way, though not always the same issues or in the same order.

In the context of looking for a career in environmental conservation in Britain and Europe today, however, the most useful analysis of the issues is the Union-wide assessment recently published by the European Environment Agency under the title *Europe's Environment*, for its findings will largely determine the design of the next Action Programme, beginning in 1996, and hence the EU legislative and financial programme for official moves affecting environmental conservation in industry and practically everything else until the year 2001.

Since 1992, following the establishment of the single market by the Union Treaty, the job market in environmental conservation available to British citizens without work permit restrictions has - in theory - expanded to embrace the whole of the EU, so a continental perspective makes sound job sense.

The assessment reports on the state of various environmental 'compartments' in a Europe-wide context, relates that state to prevailing environmental pressures and assesses different human activities as contributors to these pressures. A summary of *Europe's Environment* is available without charge from the EEA, Kongens Nytorv 6, DK-1050,

Copenhagen. Rather than expend limited space on a survey of topical issues that could only be misleadingly selective or partial, this handbook simply lists the main headings under which the assessment is arranged (see below) and recommends that those who seek further information send for a summary and follow up for themselves.

Environmental Media Assessed in Europe's Environment

Air (includes atmosphere and climate)
Inland waters
The seas
Soil
Landscapes
Nature and wildlife
The urban environment
Human health

Environmental Pressures Examined in the Assessment

Population, production and consumption
Exploitation of natural resources
Emissions
Waste
Noise and radiation
Chemicals and genetically manufactured organisms
Natural and technological hazards

Human activities Assessed in Europe's Environment

Energy
Industry
Transport
Agriculture
Forestry
Fishing and aquaculture
Tourism and recreation
Households

Specific Problem Issues Identified by the Assessment

Climate change
Stratospheric ozone depletion
Loss of biodiversity

Major accidents
Acidification
Photochemical smog
Managing freshwater resources
Coastal zone management
Forest degradation
Waste production and management
Urban stress
Chemical risk

Careers in the Independent Sector

Chapter 3

Careers With PVOs and NGOs

PVOs are private voluntary organisations - independent groups, usually charities, whose income arises mainly from membership subscriptions and whose work (mostly campaigning, lobbying and fundraising for primary environmental care initiatives) is often done through members who work for free. Some are identified with a single issue, such as conserving marine mammals, others pursue an all-in environmental agenda.

NGOs (non-governmental organisations) more often tend to be national or international institutions that depend on official funding sources, partner-donors who expect services in return, philanthropic trusts or major foundations, rather than on a mass membership. They are, on the whole, more likely to engage in technical or education programmes than in advocacy or fundraising campaigns and less likely to rely heavily on volunteer aid. Because many of them originated as voluntary organisations, they are often lumped together with PVOs as organisations in the 'voluntary sector'.

The 'voluntary sector' used to be entirely that - people gave their time *gratis* to their chosen causes. As the profile of environmental issues soared in the 1970s, all leading voluntary bodies took to employing a nucleus of professional staff and developed a salaried career structure of their own. The professionals were mostly there to carry out policies set by governing bodies representing members, to see to the administration of services and activities, and to raise more funding to start up new activities or to cover core costs.

In the early 1980s, rising unemployment led to government-funded job creation schemes. In a short time, many groups that formerly managed on a pittance could afford to double their payroll or better. This boom was over by the end of the 1980s but new expectations and an authentic career structure had been created, from which there was no going back.

Heartfelt commitment is still much valued in the independent sector but this is now bolstered by a professionalism which has done much to galvanise other sectors into action. In the past five years membership and sponsorship have dropped through recession but voluntary organisations are, on the other hand, being taken more seriously and consulted by others on everything from planning enquiries to packaging designs. It is largely thanks to this sector that there are now more jobs elsewhere in conservation, in relatively affluent organisations in the public and private economic sectors. Much that was previously done by voluntary bodies has also become a mainstream occupation for local authorities, official agencies, quangos and industrial developers. What has been lost in this process, however, is the popular participation element that made the first PVOs so effective.

In the 1990s, PVOs and NGOs have been finding new and better ways of playing to their strengths, for example through strategic partnerships with others that can complement and reinforce their impact on events and through making the most of the participatory, 'bottom-up' nature of things they do. They also seek increasingly to combine fundraising, education and campaigning activities into a unity, so that (for example) a campaign stunt also becomes an occasion for recruiting new members, or a fundraising campaign also serves educational purposes. This is a more ingenious and calculating process than many outsiders or beginners might imagine.

Starters seeking a career in this sector can still just about get by with a creative urge to dedicate themselves to pro-environment ideas and values, or (on the other hand) a thorough technical understanding of a specialised aspect of environmental science and problem-solving. But the ideal aspirant has both, and more. For unquestionably the master-key to success in this sector is *versatility*. Even in the most dedicated campaigning organisation, the first requirement will be for recruits who can do an ordinary job well - excellent accountants, secretaries, planners, researchers and so on.

Employers also increasingly expect relevant formal qualifications. Many UK universities have responded to rising demand for 'green-collar' qualifications by adding 'greened' components to existing courses or developing totally new core curricula in environmental sciences, or both (see Chapter 12).

The best qualifications for careers in PVOs and NGOs are, generally speaking, the most interdisciplinary. A degree or HND in old-fashioned Geography or Economics can sometimes open more doors than a more technology-specific qualification in Environmental Sciences, while a broad Biology or Life Sciences qualification is just as acceptable to most employers as a more specialised degree or diploma in, say, Ecology. They also place more fallback careers at the job-seeker's disposal.

In some jobs, it is still true to say that academic qualifications, while they might help settle a 'tie-break', are not as important to the employer as a keen sense of vocation and a solid record of practical experience.

Yet how to acquire experience in the first place? Doesn't the usual Catch-22 operate, the one that decrees that experience can't be acquired without a job, yet neither can a job be applied for without experience? Not necessarily. It is still surprisingly easy to gain valuable experience by taking advantage of openings for voluntary support work that abound in this sector or by joining vocational training schemes.

Work Experience and In-service Training

Award schemes like Duke of Edinburgh's or Prince's Trust, holiday ventures like the BTCV Natural Breaks programme, city projects run by the Groundwork Trust and by urban wildlife groups; all offer scope for respectable work experience.

Many other groups, including WWF, the RSPB, FoE and local Wildlife Trusts (contact the national Wildlife Trusts office) provide work experience opportunities through local groups. Anyone over 16 can apply to be a voluntary warden on various RSPB reserves around the country. From 20, you can apply to be a Summer Warden, a paid job that can lead to permanent posts.

A way to get qualifications and experience at the same time is with a countryside management training scheme leading to a National Vocational Qualification (NVQ, or in Scotland SVQ). The Countryside Commission, National Trust and BTCV are among the institutions that back this scheme. Seek details from the National Council for Vocational Qualifications.

If you have been unemployed for six months or more, you are entitled to join the Training for Work scheme: it helps graduates who want to work in conservation add particular skills to their repertoire. If you have been jobless for 12 months or more, you can join the Community Action Scheme. Details of both schemes are obtainable at local Jobcentres.

For field courses, distance learning opportunities and other alternative routes to education and training, see Chapter 12.

Openings and Opportunities

These vary according to the organisation. Before applying for work you should know the answers to certain basic questions. What kind of organisation is it? What sort of work does it do? How many people does it employ? Does it have regional offices? Does it have openings for people with your particular qualifications? A letter (send an sae for a reply) or a phone call will usually provide the answers. Often the best chance of gaining a permanent job is through working for the organisation as a volunteer.

Organisations vary in the way they are managed and structured, but usually include the kinds of department described below.

Administration

The bulk of the day-to-day operation of voluntary bodies consists of administration, much of which is routine. Suitably qualified senior managers deal with personnel, financial matters and office management, and are as often appointed for their management skills as for their commitment to conservation.

Working as a secretary is a useful entry into the profession, giving the enterprising person the chance to learn about conservation from the inside, and compete on a strong footing for internal promotion when it arises.

Conservation and Primary Care

Voluntary bodies nearly always employ staff to direct, manage, and carry out their conservation programme. Staff with a background in life or earth sciences are employed to determine conservation policy, manage projects, allocate budgets, advise on campaign planning, provide public information, carry out research and surveys, carry out species protection duties, give advice on educational programmes etc.

Those organisations that buy and own land usually employ salaried and voluntary *wardens* and other staff to manage their reserves. The work may involve breeding threatened species, rehabilitating hurt or injured animals, assessing the threats posed to species or habitats, monitoring the application of environmental laws, or simply running and maintaining a reserve. Wardens need not be scientists, but they should have a proven interest in natural history and countryside management, and enjoy the outdoor life. Basic rates of pay are not glamorous but perks such as housing or transport are quite often included. Vacancies for full-time wardens are always hotly competed for.

Scientific officers usually need relevant qualifications and proven ability to apply their knowledge. Campaigning voluntary bodies usually also employ staff with relevant expertise and experience to present their case to national and local government, statutory bodies, industry, landowners, and any other body whose activities affect the environment. Scientists are usually chosen for these posts, but lawyers are sometimes also needed.

Fundraising

Every employee of a fundraising body is helping to raise money in one way or another, but there are usually employees whose sole purpose is to raise money by specific means.

Some charities use commercial promotions where businesses lend their names by advertising the charity and donating a percentage of sales to the organisation. Other charities make direct appeals for donations to commerce, industry, and charitable foundations. Encouraging deeds of covenant from members and prevailing on people to leave money in their wills are two other methods of raising money which are widely employed.

Many voluntary bodies run trading systems whereby they sell products, which are often specially commissioned (for instance, with a logo or natural history theme), either through a network of shops or through a trading catalogue. The profits from trading are then put into conservation. Signing up members is another form of fundraising which also encourages active involvement in the activities of the charity.

Candidates for fundraising work should normally have proven commercial experience in fields such as marketing, sales and promotion.

PR and Media Relations

As voluntary bodies rely almost entirely upon public support, public awareness is crucial to their survival. Almost every voluntary body employs at least one or two people, and sometimes a whole department, to process public inquiries, generate press coverage, attract free advertising, write and publish information material, and arrange conferences and exhibitions. Previous experience in advertising, public relations, journalism or information science is an advantage.

Education
Many of the larger voluntary bodies run youth membership groups and education programmes designed to involve teachers, youth leaders, schoolchildren, and youth groups in practical conservation or fundraising. Increasing effort is being put into making sure that conservation is part of school curricula, and conservation education officers try to promote an interest in natural history in children by the use of lectures, film shows and field trips. A desire to work with children is a prerequisite, and teaching experience is usually required.

Regional Staff
Regional offices are run only by the larger voluntary bodies, and can offer openings for people with no experience in conservation although various organisational skills are required. Regional offices are usually smaller versions of the national headquarters, with a small group of people promoting the activities of the body at the local level. This could be fundraising, practical conservation, reserve management, education, or conservation.

To understand the structure of voluntary bodies and to gauge the types of opening available, it is worth looking at a few organisations in more detail. Bear in mind, though, that there are many very different bodies from the ones listed below.

WWF (World Wide Fund For Nature)

WWF (World Wide Fund For Nature) is the world's largest independent conservation organisation. With its headquarters in Switzerland it has national and associate organisations in 28 countries (including Britain) as well as programme offices in a further 21 countries. WWF works with governments, industry, the media and the public to protect our threatened environment.

WWF was set up in 1961 by a small group of people, including the late Sir Peter Scott. It aims to preserve the extraordinary variety and range of life on earth; to use natural resources such as water and timber in sustainable ways and to reduce to a minimum pollution and the wasteful consumption of resources and energy.

In Britain, it has helped to safeguard over 6,000 hectares of wildlife habitats, as well as protecting native species of bats and butterflies, flowers and birds, reptiles and amphibians. The organisation has also given money in the form of grants or loans to some 250 other United Kingdom conservation organisations so that they can carry out their vital specialist work. For example, with WWF grants the Woodland Trust has purchased around 45 sites in nearly every county of England and Wales. WWF supports some 150 new conservation projects in Britain each year.

In the UK it employs a staff of 175 in its head office in Godalming, Surrey and has a network of 17 regional staff. An office in Aberfeldy, Scotland employs six. While full-time openings are limited, there are unrestricted opportunities for voluntary fundraising work with over 300 local supporters' groups and independent fund-raisers.

WWF divides its activities into four main areas: Conservation Programmes, Education and Awareness, Marketing, and Finance and Services.

Conservation Programmes. This department consists of managers, scientific officers, secretaries and/or assistants who work on specific issues and policies including: forests, marine, pollution, wildlife law, development policy, industry, managing international, European and UK projects, planning and the UK countryside, etc.

Education and Awareness. This department consists of senior staff, education, teacher liaison and publishing officers and their assistants. They are responsible for initiatives to encourage and assist schools and colleges, support professional and occupational training and community and family education as well as environmental education initiatives worldwide.

Marketing. This department consists of a wide range of sales and support staff in attracting and maintaining WWF members and raising funds. Areas include: database services which maintains 1.8 million supporter records, legacies, fundraising from government sources, trusts, mail order trading, licensing and sales promotions, corporate partnerships. A team of regional staff look after a network of volunteer supporter groups.

Finance and services. This department contains professional and support staff and deals with accounts, personnel, office management, computer services and other administrative jobs. In addition, there is a press office and a publication section which contains writers/editors, designers and an information officer, as well as a library and photolibrary.

Friends of the Earth (FoE)

Friends of the Earth is one of the leading UK and international environmental pressure groups and campaigns on a wide range of issues to bring about changes in policy and practice and raise public awareness. It also seeks to empower individuals and community groups to play a role in helping protect their environment. FoE produces authoritative research which is used by commerce, governments and other environmental organisations and also produces a wide range of public information publications.

Founded in 1969 in the United States, FoE now has national branches in 56 countries which make up FoE International, the largest network of national environmental groups in the world. The British branch of FoE was founded in 1971. FoE has a unique network of over 250 local groups throughout England, Wales and Northern Ireland. Local groups are autonomous and are staffed entirely by volunteers. Local groups get involved in both local and national campaigning activities. For further details of your nearest local group contact the FoE switchboard.

FoE currently divides its work into the following campaign areas:

Biodiversity and habitats. Species protection, forests (including rainforests), UK wildlife habitats.

Energy, nuclear and climate change. Nuclear, fossil fuels, renewables, energy usage and climate change.

Industry and pollution. Industrial emissions, water pollution, waste and land contamination.

Atmosphere and transport. National transport issues, road building, vehicle emissions, air quality, ozone depletion and acid rain.

FoE employs over 100 staff, based mainly at its head office in north London, but also has a small number of staff based in Luton, Birmingham, Bristol, Brighton, Cambridge, Belfast, Cardiff and Sheffield.

The organisation is split into five departments:

Directorate. This department consists of the Director's office and the Information Technology team.

Campaigns. The Campaigning Department consists of four campaign issues teams (covering the issues outlined above), a small research unit working on sustainable development and international issues and a Local Campaigns Unit which develops and co-ordinates local campaigning and provides support to FoE's 250 local groups.

Publications and information. This department is split into four teams: *Enquiries* - dealing with enquiries from the public, *Information* - responsible for press and publicity work, *Publications* - overseeing the production and marketing of all FoE publications and *Art and Design* - providing an in-house design service to the organisation.

Finance, administration and personnel. This department manages the finances, office management and personnel functions at FoE. It also includes a Supporter Services Unit (based in Luton) which co-ordinates membership processing and the despatch of mail order publications.

Fundraising. Four teams of fundraisers cover trading, appeals, major donors and events and local fundraising.

Volunteering Opportunities

FoE has over 40 volunteers working mainly in its head office in London. Volunteers play a very important role in enabling FoE to carry out its work. If you are interested in pursuing a career in environmental campaigning or administration within the voluntary/environmental sector, volunteering at FoE offers an opportunity to gain valuable work experience. Volunteers carry out a variety of administrative support work from helping with mailouts and sorting incoming post to assisting with research and information gathering work. For further details about volunteering opportunities, contact the Volunteer Co-ordinator at the London office.

Employment Opportunities

FoE employs staff in a wide range of roles; campaigning, research, fundraising, finance, administration, IT, publicity, etc. FoE is an equal opportunities employer and all permanent vacancies are advertised externally, mainly in the *Guardian* on Wednesday. Posts which require specific technical or specialist knowledge and skills, eg IT, research or fundraising posts are generally advertised in the relevant specialist journal. If you are interested in working at FoE, plese do not send in speculative enquiries (sadly they cannot offer to keep your CV on file). Please look out for ads in the national press and apply for vacancies as they are advertised.

Further information is available from FoE's own *Annual Review*, *Publications Catalogue*, *Membership Leaflet* and *Earth Matters* (FoE's quarterly supporters' magazine).

The Centre for Alternative Technology

By concentrating on the display of working examples, by educational work, and by the provision of information services, the Centre for Alternative Technology near Machynlleth in Mid-Wales aims to influence people's attitudes towards technology and their own life-style in favour of practices which conserve and protect our planet.

Founded in 1974 and supported by the Alternative Technology Association, the Centre has grown to be a major visitor centre, and now attracts over 85,000 people each year. Several hundreds more, including specialist groups of teachers, architects and others, attend courses on wind power, water power, green living and teaching green.

Some of the topics covered by the exhibits and skills at the Centre include wind, solar, biomass and water power, organic horticulture, low-energy building, and electronics for Third World and environmentally sensitive use. Additionally, the Centre runs a bookshop - widely regarded as the most comprehensive alternative technology bookshop in Europe - and a restaurant.

Staff numbers are around 30 permanent workers, plus temporary staff. Additionally, opportunities exist for volunteers to spend periods of a week or six months working with the staff. Write to the Volunteer Coordinator. Current permanent staff employed at the Centre include engineers, builders, bookshop and restaurant staff, gardeners, publicity and information officers, display and graphic artists, and people to organise courses and visits, and handle finance and general administration.

As part of their New Futures series of environmental technology publications, the Centre has released a book to help people find a career that does not involve harming the environment. It is called *Careers in Sustainable Technology* and is available from the Centre.

Greenpeace

Founded in 1975, Greenpeace campaigns vigorously to protect endangered species and the environment. It has sent its boats into dangerous situations to prevent the killing of whales and seals, nuclear testing in the Pacific and the dumping of toxic waste at sea; it is currently campaigning against global warming, depletion of the ozone layer and air pollution as part of the international Atmosphere and Energy campaign. Salaried posts are advertised in the *Guardian*, specialist journals relevant to particular positions and occasionally other publications. Volunteers for work in the London office are encouraged to apply to the reception staff. Greenpeace also has a network of local support groups.

Marine Conservation Society

The only UK charity dedicated to the protection and conservation of the marine environment for wildlife and future generations, the Marine Conservation Society researches and campaigns on marine pollution, habitat and species conservation and the sustainable use of marine resources. The MCS publishes the *Good Beach Guide* and factsheets on subjects including sharks, sewage pollution, careers and factpacks on Marine Pollution, Marine Habitats and Marine Species. In 1993 the Society launched Beachwatch, a campaign to raise awareness about marine debris involving volunteers in an annual national beach clean. A network of regional local groups have been established to involve members and the public in campaigns and surveys. Marine conservation study packs for schools have been produced for primary and secondary level. The Society's headquarters is at Ross-on-Wye, Herefordshire; staff numbers are small; voluntary helpers become involved with projects such as Seascarch (underwater survey work calls for experienced divers), Beachwatch, basking shark and marine wildlife and pollution surveys.

British Trust for Conservation Volunteers (BTCV)

BTCV involves individuals and communities in practical environmental projects. From working only on national nature reserves, its remit has now expanded to include education and amenity work in the countryside. Each year over 84,000 volunteers are equipped and trained to carry out environmental work on sites across the countryside, with projects ranging from the protection of wildlife habitats to the improvement of access to the countryside. Opportunities for voluntary work in the environmental field include spending between 3 and 12 months working as a voluntary field officer in one of BTCV's residential centres, volunteer involvement on a conservation working holiday, training as a leader to work on residential conservation projects and working at weekends with affiliated local groups (over 600). BTCV is one of the partners in UK2000, involving people of all ages in practical conservation, from drystone walling to nature reserve management, pond clearing to working in tree nurseries. BTCV Enterprises, a subsidiary of BTCV, offers products and services and develops enterprise in environmental projects.

BTCV has a permanent staff of over 195, working in around 90 offices throughout the country. Most work at regional or local level, promoting the involvement of volunteers in the projects. The most important position is that of *field officer*, with a range of responsibilities according to locality: creating a wildlife garden in an inner-city area, or dealing with the problems of tourist pressures on footpaths in a National Park. Their work includes setting up training courses, organising working holidays, running projects for unemployed or retired people and involving schoolchildren in environmental work. Most field officers have a degree background in environmental studies and several years' experience in practical conservation work, often as a BTCV volunteer.

BTCV's counterpart organisation in Scotland is Scottish Conservation Projects (SCP).

RSNC: The Wildlife Trusts Partnership

Also a UK2000 partner and formerly known as the Royal Society for Nature Conservation, the Wildlife Trusts Partnership is concerned with all aspects of wildlife protection. It is a partnership of 47 Wildlife Trusts, 50 Urban Wildlife Groups, and WATCH, the junior wing. Together they protect over 2000 sites covering more than 56,000 hectares. The partnership has a total of 500 full- and part-time staff. Each Trust has between 4 and 40 permanent members of staff. As well as graduates with degrees in subjects such as biological sciences, ecology and conservation, and those with experience in voluntary work, there are opportunities for those with PR and marketing skills.

Other voluntary organisations include:

The Council for the Protection of Rural England (CPRE) works for a living and beautiful countryside. It has 43 county branches, 160 district committees and *ad hoc* campaign groups.

The National Trust, which protects places of natural beauty or historical interest, and employs many specialists with professional qualifications and experience and also gives Youth Training to school leavers in amenity horticulture, forestry and countryside management. The Trust provides training opportunities for around 36 school leavers a year through its 16 Regional Offices in England, Wales and Northern Ireland. Specialist staff include land agents; archaeological, forestry, nature conservation and horticultural advisory staff; gardeners, forestry staff and wardens; architects, building and conservation staff, accountants, personnel managers, advisers (on art, history, textiles, paper, architecture, etc), fundraisers, caterers, retail managers and many others. The Trust also has a volunteers section, devoted to managing the activities of 28,000 volunteers nationwide.

The Civic Trust is concerned with improving towns and cities, and the economic and social regeneration of urban areas through conservation. It is an umbrella organisation for nearly 1,000 local Civic Trusts.

Case Studies

Alan works in the regional office of a national conservation charity.

> I actually discovered conservation by accident. I read geography at university, not knowing what I wanted to do after graduation, and during a vacation a friend of mine took me along to a two-week work camp on a nature reserve. I suppose I'd always been sympathetic to the aims of conservation, but my interest was really aroused in those two weeks by meeting and talking to people working in conservation. I subsequently worked on several reserves and conservation projects at weekends and during holidays, and on one project met the regional officer who offered me a job. The fact that I had practical conservation experience and knew a fair bit about the charity I now work for were probably the deciding factors, although the geography degree was useful.

At first Alan was given largely routine and menial duties, but then the regional officer moved on to the head office, and Alan was offered his present job.

> It's a good job for the time being, but I can't see myself staying here for ever. The regional office is a miniature version of head office, the difference being that I'm involved in every aspect of the job and I'm effectively my own boss. If anything, you tend to be completely forgotten by head office. The work is mostly fund raising, generating local publicity, and giving talks to schools and societies. It means a lot of travel, a lot of work with the public, acting as a one-man information service, and even doing the occasional newspaper or radio interview. The pay isn't that good and the hours are long, but there is a lot of satisfaction in seeing the amount of money you're raising and watching your membership figures climbing slowly.

Alan's future plans are to pick up as much experience as possible and then move up to head office (or another charity) and become more involved in information and conservation policy.

Fiona works in the national office of an environmental pressure group. She left school when she was 16 and did a variety of jobs before taking a secretarial course and moving to London to work as a secretary. After two years of temping she was sent to the pressure group for a one-month stint that turned into a permanent job.

> I took a big drop in salary, but I'd heard a lot about the group and got very interested in what it was doing. Most of the time I'm a jack of all trades - we all are - but I work mostly on the publications side. The nicest thing about working with the group is that there is no real hierarchy and everyone shares the work. In the past year I have helped organise a rally in Hyde Park and a pop concert, helped start up five glass and paper recycling schemes, and I've just been put in charge of liaising with the various celebrities who support us. I'm learning a lot of skills as I go along - you just have to, because you have to pay your own way and not be a passenger. The biggest problems are the low pay, the erratic hours,

and the cramped office space. You also sometimes feel you are a bit on the fringes of the real world, but the sense of doing something socially useful makes up for all that.

Chapter 4

Careers in Environmental and Resource Management

Forestry

Forestry is the art and science of managing forests, which includes everything from raising seedlings and transplants in nurseries, to felling and transporting wood, planting and tending new forest plantations, managing woodland for recreation and amenity, and conserving the forest environment and its wildlife.

Forests are a renewable natural resource and, as such, need careful planning and management. Trees not only provide timber for everything from paper to furniture, building material, fuel and fencing, but also have a vital ecological role. They provide a natural habitat for plant and animal life and have an important bearing on soil cover, water retention, wind and climate. New plantings of coniferous trees make more allowance for wildlife and conservation of landscape than previous plantations, with less wholesale cutting down of large areas, a greater mix of trees of different ages, and even natural regeneration by wind-carried seed, plus unplanted areas as wildlife habitat. Trees are an important feature of the landscape, particularly in towns where they help to soften the built environment.

The art of raising and tending trees is old, but the science of forest management is relatively new. It involves planning, predicting timber demands, controlling felling and replanting, ecological management and marketing. It is central to national resource policy. For instance, Britain was once almost entirely covered in woodland, but now has 8 per cent forest cover, less than any country in Europe except Ireland and the Netherlands.

There are over 2 million hectares of productive woodlands in the UK, 827,000 hectares of which is managed by the Forestry Commission. The rest is divided between private forest estates, private woodlands and woodland owned by timber merchants.

What the Job Entails

Forestry at the moment employs about 23,000 people, working mainly in the Forestry Commission and the private estates. Others work in forest management companies, harvesting companies and wood processing industries. The Forestry Commission's structured employment pattern contrasts with that of the private sector, where career development depends on the individual employer.

Private estates vary in size from less than 40 to more than 200 hectares, and employ some or all of the following staff:

Land agents (*factors* in Scotland) put the policy of the owner into practice; they must be members of the Royal Institution of Chartered Surveyors or the Incorporated Society of Valuers and Auctioneers. *Forest officers* are given charge of the forestry section. *Foresters* are responsible for planning the annual programme of forestry work, supervising and training forest workers, and controlling the progress of work schemes and specialist activities such as research and wildlife conservation. *Foremen* oversee work parties. *Forest workers* (or *woodmen*) are the craftsmen and team workers, and are trained for a variety of jobs from clearing and weeding to planting, servicing and operating machinery, using chain saws, applying fertilisers, supervising nurseries and pruning and thinning trees. *Rangers* and *wardens* manage and control wildlife, maintain campsites and recreational facilities, and guide visitors around forest reserves. Other workers operate machines, run sawmills, and undertake other skilled and unskilled jobs.

Qualities Required

Most foresters are united by an interest in, and understanding of, the needs of wildlife and the countryside, and of the forestry industry. They live mainly in the country or small towns and villages, and must enjoy the outdoor life, be able to work alone while fitting into the needs of the industry at large, and be able to express and communicate the essentials of forestry to administrators, planners, farmers, sportsmen and the public.

Qualifications

These vary according to the type of work, but all except forest workers and woodmen need specialist training. This could be basic craft training for new entrants provided by the Forestry Commission and Forestry and Arboricultural Safety and Training Council (FASTCO), or a diploma or degree in forest science and management.

Openings and Prospects

The number of people employed in forestry has decreased, partly because of the introduction of new methods of making the industry more efficient and productive in the face of overseas competition. However, numbers have now stabilised.

In the UK there are opportunities in local government service, in national parks, in voluntary land-owning bodies such as the National Trust, in commerce and industry, in teaching and research, and in private forests and with the Forestry Commission. There are also many opportunities of various kinds overseas, on short-term contracts in the Third World or in permanent posts in countries such as Canada, Australia and New Zealand.

Water Industry

The demand for water for domestic, industrial and commercial consumption has been increasing steadily in recent years, and even in a country with a normally high rainfall, such as the UK, water shortages have become more frequent. The importance of managing and conserving water resources, the need to treat waste water and control pollution, and the importance of flood protection led to a review of the organisation of the water industry.

The main employers in England and Wales are now 10 private water service companies and the National Rivers Authority (see Chapter 5).

In Scotland, there is a different two-tier system with nine regional councils divided into 53 district councils. Purification is handled by seven area groups responsible for the quantity and quality of water discharged into rivers and coastal waters.

What the Job Entails

Like all natural resources, water is governed by limits, and careless use will lead to shortages and deterioration in the quality of rivers, lakes, dams and reservoirs. Water often needs to be recycled to bolster supply, which involves treatment to remove effluent and sewage.

RWAs and their Scottish equivalents are responsible for forecasting water demand, developing new sources, flood forecasting, hydrometry, storing and supplying water, land drainage, and fisheries. They are also responsible for the conservation of water as a resource, the control of river pollution, and nature conservation. They employ biologists and geologists, specialists in fisheries management, civil engineers, hydrologists, and a variety of administrative staff and skilled and unskilled workers.

Openings and Prospects

Although openings for biologists and water pollution control scientists are still fairly limited, the RWAs are showing more interest in graduate recruitment. There are also openings in advisory, supervisory, technical and managerial positions in water resources, distribution and supply and treatment areas.

Energy Industry

Energy is the backbone of our industrial and technological way of life. It provides light, warmth and power, cooks our food, fuels our transport systems and industries and, above all, makes the existence of life on earth possible. Over the centuries we have generated energy from a wide variety of sources (from charcoal to nuclear power) and have created extensive industries to extract and manage our energy needs.

At the moment all but 8 per cent of our energy comes from three sources: oil, coal and natural gas. They are all fossil fuels and are all non-renewable. This means that one day they will run out. The result is that we are now beginning to use energy more efficiently and intensively by cutting down on waste, and are also investigating new sources such as solar, wind, wave, bioenergy and geothermal energy.

Bioenergy covers a variety of technologies, including the production of methane from waste and the growing of energy-crops, and is the sector expected to expand the most.

Large-scale wind turbines and wind farms linked to a national grid have been developed by large construction companies, and by GEC and British Aerospace, and the construction of tidal barrages is being considered as a way of generating electricity. It is also possible that shipyards could be used to manufacture wave energy converters. Large firms such as Calor, Pilkington and Philips are also interested in wave power and solar energy. Solar power, bioenergy, energy conservation and heat pumps are all in the research and development stage, both in university research groups and with smaller engineering companies, many of whom are already manufacturing water turbines and small wind turbines for domestic and agricultural use.

Energy is a growth area for jobs, with 5,000 to 10,000 jobs in the renewable energy field, including hydroelectric plants, plus energy conservation work. Energy managers are employed by both industrial companies and local authorities to reduce the use of energy; it is their job to make cash savings, challenge accepted practices and propose feasible alternatives, and prevent detrimental effects on the environment. A list of manufacturers and installers of renewable energy equipment is available from the Centre for Alternative Technology.

Warwick University has an AT-specific degree course: Engineering Design for Appropriate Technology (EDAT). Some science-based degrees contain AT options, such as the Society and Technology degree at Middlesex University, and special energy or environmental degrees at colleges, including Napier University, Edinburgh, the University of East London, and Reading University. Many of those employed in the field of renewable energy are architects or engineers. Information about suitable energy courses and jobs is given in the NATTA (Network for Alternative Technology and Technology Assessment) newsletter; details are available from NATTA Energy and Environmental Research Unit, Faculty of Technology, The Open University, Milton Keynes MK7 6AA.

Landscape Architecture

Landscape architecture is a small, young, but fast-growing profession with about 3,400 professionally qualified members and the demand exists for many more. The career offers excellent prospects and considerable job satisfaction to anyone with an interest in the natural environment.

Landscape architecture consists of planning and designing open spaces. Landscape architects can both protect existing features and help nature to return to otherwise sterile urban locations.

What the Job Entails

The landscape architect analyses and resolves the demands made on open spaces, and uses the spaces to best effect. This might involve restoring derelict land, or landscaping a new town scheme, a housing or industrial estate, shopping precinct, golf course, marina, cemetery, a stretch of motorway, a public park, or a private garden. Each design has to consider the ecological, functional, aesthetic and management aspects of the scheme.

The landscape architect usually takes a brief from a client, carries out a detailed survey of the site, and then designs a scheme that draws on the advice of ecologists, planners, architects and engineers. Once a contractor has been appointed, the landscape architect works on the site to oversee the implementation of the scheme.

There are three specialist areas within the profession:

Landscape architects, trained in design as well as technical skills, supervise the scheme on the drawing board and on site. *Landscape scientists* are especially concerned with environmental factors and have qualifications in a natural science plus expertise in ecology, soil science, geology, hydrology or plant or animal biology. *Landscape managers* have a background in horticulture, forestry or agriculture and are responsible for the long-term care, development and management of the scheme. Landscape constructors may also be on site. Their skills range from unskilled manual tasks to skilled management, and combine a knowledge of horticulture and agriculture with basic building practice.

Qualities Required

Landscape architects should have an instinct to preserve and conserve, and should be deeply concerned about the natural environment. The job demands an interest in both art and science, and in people and the way they relate to the natural environment. Landscape architects also need to be physically fit, and prepared to spend most of their time out of doors.

Qualifications

Membership of the Landscape Institute is the recognised professional qualification for landscape architects in the UK. To be eligible for associate membership of the Institute, candidates must have two years' work experience and pass the Institute's professional practice exam, having successfully completed a course at one of the recognised schools of landscape architecture in landscape architecture or landscape management. There are no specific courses in landscape science, but some courses, such as the MSc in Conservation at University College London and the MSc in Ecology at Aberdeen University, are appropriate.

Openings and Prospects

Landscape architects work either in private practice or for government, industry, development corporations, or planning departments, and are commissioned to work both on private and public land. As the profession grows, so the number of openings will increase. Private practice currently offers most of the opportunities, but government and public and private organisations are taking on more landscape architects. There are also openings overseas, either working on projects given to British consultancies or working with overseas consultancies.

Surveying

The surveying profession offers a diversity of career opportunities to graduates from all disciplines.

Chartered surveyors are the property professionals instrumental in developing and maintaining the very visible environments which affect all our lives.

Surveyors are involved in the full life cycle of property from its conception to planning, development, valuing, marketing, managing and refurbishment. Contrary to popular conceptions of the role of chartered surveyors, their expertise is not confined to the construction site or the estate agency. Surveyors may also advise on environmental issues, rural conservation, historic buildings, mineral deposits, geographic information, marine resources and the valuation and sale of livestock, fine art and antiques.

This presents an incredible array of career options. In order to pursue a career within such a diverse profession you will clearly need to specialise but whichever area you choose you will learn a whole range of marketable skills and be professionally qualified and trained. The precise nature of your work will depend on the discipline or disciplines you choose to practise in and the type of organisation you work for.

A career as a chartered surveyor could involve you with employers ranging from government departments, to local authorities, nationalised industries, commercial companies, building companies, private landowners, private practices and auctioneers.

Qualities Required

Surveyors are members of a profession that demands a high degree of social responsibility. They need the knowledge and breadth of outlook to understand how resources need developing and planning, and the flair for seeing and understanding all the social, aesthetic, financial, economic and legal aspects of the job. This means working as part of a team of experts in different fields, and being able to contribute to collective decisions. Land and hydrographic surveying demand a high degree of technical and mathematical ability.

Qualifications

Employers actively recruit graduates from a variety of backgrounds because, primarily, surveying is concerned with communication skills and business acumen.

There are two components to qualifying as a chartered surveyor. First you must successfully complete a degree or diploma accredited by the Royal Institution of Chartered Surveyors (RICS) and second, a period of practical training known as the Assessment of Professional Competence (APC).

Postgraduate conversion courses are also available; full-time one year and part-time two years.

The variety of accredited degree and diploma courses equip the graduate with multi-disciplinary skills including law, management and economics and so provide the ideal stepping stone to any career in property or business. To pursue a career in property you will be required to complete the period of practical training (APC) while you work.

The Assessment of Professional Competence requires the graduate to record their daily work in a diary and logbook and successful completion leads to chartered status. The minimum period for the student to qualify, including the academic and practical study, is three years for the postgraduate route and five years for the first degree route.

Whichever route you decide to take you will have a career ahead of you that may include work in the country, at sea or in the city. As a chartered surveyor you are not even restricted on territorial grounds because your qualification enables you to practise your skills anywhere in the world.

Openings and Prospects

Further information on chartered surveying companies is available in the *RICS Geographical Directory*. The RICS careers brochure has information on all surveying disciplines and a list of RICS accredited postgraduate courses.

Parks and Recreation

As an increasing number of people have more leisure time, so the demand for better recreational facilities grows, and both national and local authorities are putting more energy into meeting this demand. Working in parks and recreation involves providing a variety of outdoor and indoor facilities, most of which are designed to increase people's enjoyment of natural features. The work might mean managing and planning local town or city parks, woodland, estates, playing fields, cemeteries, crematoria, allotments, and almost any site designed for public recreation.

An imaginative manager could turn a barren piece of land, in an unprepossessing urban area, into a haven for wildlife, or carefully convert a cemetery into an attractive local nature reserve. The job incorporates aspects of horticulture, arboriculture, landscape design, construction and conservation.

The 11 national parks in England and Wales, including the Broads, are privately owned, mainly by farmers and landowners and the people who live and work within the park areas, and administered by local boards and county councils. Farmers and landowners are encouraged and helped to carry out work that contributes to the conservation of the landscape and wildlife. Landowners include the National Trust, and there are also some areas managed by the Forestry Commission and wildlife habitats protected by voluntary organisations such as English Nature.

Conservation in the parks is the responsibility of park rangers, or wardens, with help from volunteers - who may use this as a route into a permanent job.

The members of the Association of Countryside Rangers, who work in the National Parks and other conservation areas, are involved in the daily management and conservation of sites and larger areas, and in environmental education (including guided walks for visitors). Over half work for local authorities; other employers, apart from the national parks authorities, are the National Trust, the Woodland Trust, county naturalists trusts and English Nature.

What the Job Entails

There are three levels of career in park and recreation management: professional, technical and craft. All are interrelated, and there are possibilities for promotion within and between the different levels.

Craft work demands a variety of different skills that can be learned, either on the job or on a three-year course that involves working and taking day-release classes at colleges of further education. The work involves anything from using and maintaining tools and machinery to laying out horticultural features, cultivating plants, maintaining trees, building paths and drains and constructing special public facilities.

Technical work involves the same duties at a more specialised level, and *professional* posts are reserved for those with appropriate qualifications and management abilities.

Qualities Required

A love of, and understanding for, the needs of nature are essential, as is an interest in meeting the recreational needs of the community. Members of the profession work as a team to meet those needs and to maintain minimum standards on each scheme. An interest in horticulture (the cultivation of plants), arboriculture (the cultivation of trees) and conservation are all part of the job.

Qualifications

The body that promotes and oversees the profession is the Institute of Leisure and Amenity Management, which promotes education in the field and sets professional standards. Membership of the Institute is the recognised professional qualification. Technical posts call for candidates with four (appropriate) GCSEs, or at least one science A level, who may be eligible for specialised training in amenity horticulture or a technical subject such as landscape and horticultural technology.

Candidates for professional posts usually need at least four GCSEs and one A level, and to spend time as trainees attending day-release classes, followed by further training in amenity horticulture. Candidates with two or more A levels may go on to read horticulture at university and work in the service during vacations.

Openings and Prospects

As most jobs in park and recreation management are currently offered by local authorities responsible for parks, recreation, amenity and leisure, so opportunities vary with the policies and budgets of each authority. There are also occasional vacancies with the national parks authorities. Although work for local authorities is subject to competitive tendering, most of the work is carried out by the authorities' own recreation and leisure departments, rather than by private contractors.

Working in Industry

Industrial technology gave us the means to exploit the earth's resources over the past 200 years. However, many environmental problems have been caused by unrestrained and unplanned industrial development. Perhaps the most obvious bad side-effect has been pollution.

The situation is now changing: the tightening up of legislation, the requirements of central and local government, and a greater awareness of the need for minimum standards of environmental health have combined to make industry take its environmental obligations more seriously in recent years.

While there has been a rise in the number of environmental planners and scientists employed by industry, openings are still limited and it is difficult to make any generalisations or firm forecasts. Many industries have been able to 'clean up' by tightening up their existing policies and by making their workforces aware of the issues. They have not found a need to employ environmental specialists. Other industries have employed ecologists and biologists to monitor anti-pollution measures, the effects of industry on local ecosystems, and the needs of resource planning. Much of this is supported by intensified public relations programmes.

One company might be looking for extremely well-qualified specialists, such as biochemists, plant ecologists, microbiologists, or botanists, to work on specific projects. Another may do no more than consider applicants with environmental science qualifications on a par with other general applicants. Yet another company may give preference to environmental scientists for general vacancies where their expertise could be a bonus.

Because there are no hard and fast rules, looking for openings in industry is a matter of assessing which organisations are most likely to employ environmentalists, and to approach each one individually. The number of openings in this area is likely to increase especially for those qualified in chemistry, botany and biology.

There are also opportunities in large companies and consultancy groups for those with experience and qualifications in ecology and the environment, to assess and advise on the impact of projects such as civil engineering and construction works (see Consultancy below).

Working in Agriculture

Although agricultural policy has enormous influence on the quality of our natural environment, it offers few openings for conservationists and environmental planners. Nevertheless, this apparent lack of opportunity has not stopped suitably qualified environmental scientists and planners from working in agriculture, although they have to compete for vacancies with agricultural scientists. Experience in farm work, and qualifications in soil science, biology, horticulture or a similar area, are normal requirements for a job in this field. Agricultural research is another likely avenue for those with relevant specialist skills, and a few opportunities exist for work on the environmental consequences of agricultural policy and research into more efficient and ecologically sound agricultural methods.

For those without skills or qualifications, manual work on farms leading to farm management work offers the chance to supervise and plan the running of a farm, and to make decisions on how the land is used.

Under the Countryside Stewardship Scheme, developed with the Countryside Commission, English Nature, and English Heritage, farmers and other landowners receive incentives to help them combine commercial land management with conservation; the scheme has five target landscapes of chalk and limestone grassland, lowland heath, waterside landscapes, coastal areas and uplands. Farmers can also apply for the Countryside Premium, which provides incentives to farmers to manage set-aside land for the benefit of wildlife and the landscape.

The Farming and Wildlife Advisory Group employs approximately 45 Farm Conservation Advisers who give advice to farmers on: whole-farm and partial conservation plans; conserving or planting hedges and hedgerow trees; making and maintaining ponds; improving field margins, tracks and roadside verges; conserving wetlands, scrub, rough grass and old woodland; planting trees in new woodland and tree groups. They also give talks, run courses and guided farm walks and attend county shows, meetings and conferences.

One area of agriculture that is growing is organic farming, based on crop rotation, non-chemical pest and disease control, the recycling of nutrients and a healthy soil structure - as an alternative to intensive, chemically based production methods. The total acreage under organic production is increasing at around 5 per cent per month.

The British Organic Farmers, the Organic Growers Association and the Soil Association produce technical booklets and share a library and resource centre at the Organic Food and Farming Centre, 86 Colston Street, Bristol BS1 5BB. Some agricultural colleges now have short course modules dealing with organic farming and non-intensive methods of food production within existing courses. The Soil Association also provides information about education and training opportunities in organic agriculture; a list of courses approved by the Association is available from the Organic Food and Farming Association, address as above (send sae).

Other organisations involved in organic farming are the Organic Advisory Service, Elm Farm Research Centre, Hamstead Marshall, Nr

Newbury, Berkshire RG15 0HR; the Bio-Dynamic Agricultural Association, Woodman Lane, Clent, Stourbridge, West Midlands DY9 9PX; the Henry Doubleday Research Association, Ryton-on-Dunsmore, Coventry CV8 3LG, and the Centre for Alternative Technology. For practical experience, Working Weekends on Organic Farms arrange work in exchange for meals and accommodation; details from WWOOF, 19 Bradford Road, Lewes, Sussex BN7 1RB.

A career in agriculture is worth considering, but it is up to the individual to create the openings and to acquire the right training - there are few precedents to use for guidance.

Consultancy

Nowadays, public pressure and resulting legislation or regulation mean business and industry must demonstrate environmental concern and observe ever stricter standards. Investors and developers are expected to make sure before they embark on any new scheme that they have done everything in their power to avoid or mitigate any adverse environmental effects or costs it might have. Nor are official aid and development projects exempted from these requirements.

Some companies and institutions have in-house capacity to deal with these matters but in many cases they call in the services of an environmental consultant. Consultancy is one of the most rewarding professional niches in the independent sector and it is also at present a conspicuous growth area.

Consultants are usually brought in to serve an advisory function, although sometimes they are delegated executive authority to set up a one-off activity or a 'temporary institution' to manage a specific outcome, such as an environmental survey which calls for an independent viewpoint.

Consultants may also be contracted to take control of an activity which the client does not have enough personnel or expertise of its own to handle but considers too infrequent a need to merit building capacity for in-house. Consultants are expensive, but less so than in-house payroll and overheads.

Tasks which independent consultants or consultancy partnerships typically take on, include *design or planning* activities to gauge the future behaviour or functioning of a process, product or enterprise in relation to its environment.

Examples are clean product or process design for industry, designing instrumentation for environmental management activities such as air pollution monitoring; or 'greening' management structures and office procedures in a company requiring registration under standards that can earn it an environmentally sound public image; or a 'green labelling' facility from an official watchdog scheme that puts marketing claims for 'environmentally sound' products to the test.

Environmental auditing and *environmental impact assessments* are the bread-and-butter of consultancy work in industry and increasingly in the official sector. These activities require highly specific surveying, measuring and monitoring skills. In the past, methodologies applied to these tasks have varied suspiciously, so that two or three different consultant reports on the same enterprise could be expected to come up with two or three strikingly different sets of conclusions. In recent years, determined efforts have been made by the Institution of Environmental Sciences and other bodies to set uniform and exacting standards and develop methodologies that build confidence in professional environmental auditing and assessment practice, rather as chartered accountants or surveyors are expected to live up to one code of practice.

Practitioners involved in consultancy work are generally experienced veterans with a distinguished record in applied research or resource management. Membership or Fellowship of the Institution of Environmental Sciences is a useful basic qualification but selection standards are rigorous.

Other Employment Niches

This section has by no means exhausted the range of potential careers in the independent sector. Permanent positions in research and development organisations or in education and training institutions are among the more obvious absentees from the roll-call so far but, as the activities they foster feed more or less equally into the independent sector and the official sector (which forms the framework of the next few chapters) they will be dealt with in a section of their own (see Chapters 7 and 8). Careers in the media and creative domain (Chapter 9) have been paid a similar compliment.

Types of Environmental Employer

As classified by the Institution of Environmental Sciences:

Specialist environmental consultancies
Engineering/planning consultancies
Industry
Equipment and plant suppliers
Regulatory agencies
Research and development organisations
Education and training institutions
Non-governmental organisations
Environmental management organisations
Financial organisations
International bodies

(*Environmental Careers Handbook*, 1993)

Careers in the Official Sector

Chapter 5
Working for Government

Introduction

Much progress has been made in the formulation of environmental laws, controls on planning and the protection of species and habitats. The creation of the Department of the Environment in 1970 was a significant step, and other departments have since gradually adopted new policies in favour of environmental management and protection.

This has given rise to an expansion in the number of job opportunities for planners and environmental scientists. Most of these openings fall within the normal Civil Service structure, and appointments are made by individual government departments (look for job advertisements in the press). To obtain such an opening usually means first joining the Civil Service in a general capacity, and then moving to a relevant department as soon as possible. Administrative, scientific and research posts provide the most likely options for conservationists.

Openings for School Leavers

If you have GCSEs or A levels and no specialist knowledge, clerical and executive posts provide the main point of entry. The first rung on the ladder is the *administrative assistant*, whose work involves keeping records, sorting, filing, and answering public inquiries. There are no age restrictions for entry, but two acceptable GCSE passes, including English language, are normally required or candidates can take a written test.

The next step up is the *administrative officer*, who handles the incoming correspondence of the department, helps the public, either on the phone or over the counter, and keeps records and accounts. There are no age limits for entry, which is either by promotion from administrative assistant, or by successful applications from newcomers with five acceptable GCSE passes, including English language or a written test.

Executive officer posts are the most responsible of the administrative posts, which involve the application of departmental policy, managing the work of administrative assistants and officers, and working directly with the public. There are opportunities for specialist training. An applicant needs to be under 52 years old, and have two A levels and three GCSEs, or their equivalent. One of the passes must be in English language. Promotion beyond this point is normally to *higher executive officer*; entrants need two A levels and four years' experience.

School leavers can also apply for vacancies as *assistant scientific officers* to support research and project teams. Applicants should have A levels or four acceptable GCSEs, including English language and a maths or science subject. Opportunities exist for specialisation and advancement into preferred areas of work.

All these staff are recruited locally by individual departments.

Openings for Graduates

There are opportunities in the Civil Service for graduates with both general and specialist or professional qualifications. Administration is the main point of entry for the specialists and includes policy planning, the drafting of legislation, or the management of executive programmes.

Administration trainees spend two years at that level, combining their work with training at the Civil Service College. They should have the practical intelligence needed to assess and deal with problems they encounter, be able to work in a team, and be able to express themselves clearly. Applicants should be under 26 and should have a degree with at least second class honours. Promotion is to *higher executive officer*, *Grade 7*, *Grade 6*, and so on.

For those with at least a second class degree in an appropriate specialised field, there are opportunities in research areas. *Research officers* study the impact of government policy, and provide information upon which to base future policy. The Resource and Planning Group is particularly interesting because it studies policy in relation to the allocation of resources and the environment. Applicants should normally be aged under 28 and be qualified in geography, agricultural economics, economics, economic geography, or a related field. Promotion is to *senior research officer*.

Scientists are needed for advisory services, and to carry out research and development. Applicants should have an appropriate degree or equivalent qualification.

Vacancies for graduates are usually advertised nationally by the Recruitment and Assessment Services Agency, or the relevant government department.

Relevant Government Departments

While environmental specialists may be employed on a full-time or consultancy basis at one time or another by almost all government departments, the majority of openings are in the following departments.

The Department of the Environment (DoE)
Despite its name, nature conservation and resource planning are only one part of the work of the DoE. The 'Environment' in its name refers to people's living environment, and includes areas such as housing, new towns policy, and local government.

The Department also has responsibilities in planning, development control, inner city renewal, countryside affairs, pollution control and water resources. It has two headquarters offices in London and Bristol, and nine regional offices. There are five divisions of particular interest:

Planning. The DoE oversees regional and local planning policy, and is one of the largest employers of qualified planners. The headquarters of the Planning Inspectorate is in Bristol.

Royal Parks and *Historic Royal Palaces Agency.* Staff working in the Royal Parks are employed by private contractors. They include qualified horticulturalists, gardens, birdkeepers and gamekeepers; among the 350 staff in the Royal Palaces Agency there are gardeners and specialists in conservation. The DoE is also responsible for 'listing' buildings with special architectural or historic merit and giving grants to, and monitoring, bodies such as English Heritage and voluntary organisations.

Countryside and Wildlife. The DoE has a policy responsibility for the conservation and enhancement of the countryside. This includes recreation and the resolution of possible conflicts where varied demands are made of land. The Department employs some part-time specialist staff for posts such as Wildlife Inspectors. It also sponsors and works with statutory agencies such as the Countryside Commission, English Nature and the Scottish and Welsh conservancy bodies (see Chapter 6).

Pollution control. The DoE has responsibility within central government for policy on most aspects of pollution control and co-ordinates policy on the remainder. Her Majesty's Inspectorate of Pollution, with an HQ housed within the DoE, exercises direct control over air pollution from certain processes and the handling of radioactive substances. It also advises local authorities on the exercise of their functions in respect of air pollution and the disposal of waste to land.

Along with the National Rivers Authority (see below), HMIP will form part of the new Environment Agency, which will take over the regulation of waste disposal. HMIP is also responsible for the new system of Integrated Pollution Control, controlling emissions of all kinds from the industries of fuel and power, waste disposal, minerals, chemicals, metals and other industries including paper pulp manufacture, timber processes and animal and plant material treatment. HMIP's staff is expected to grow from 260 to around 400 over the next few years to cope with its extra responsibilities. A Pollution Inspector must have an honours degree in chemistry, chemical engineering, mechanical engineering, physics, environmental science or another relevant discipline, or appropriate professional qualifications and experience.

National Rivers Authority. The NRA, which will be merged into the

Environment Agency, employs around 6,500 people and is made up of ten regions based on the river catchment areas of England and Wales. The principal functions of the authority are: water quality; water resources; flood defence; freshwater fisheries; conservation; recreation and navigation. The rivers, lakes, estuaries and coastal waters protected by the NRA include Sites of Special Scientific Interest (SSSIs) and wildlife habitats, with corridors and wilderness areas in towns and cities as well as farmland. The Authority is also responsible for conserving archaeological features and historic buildings. Ecologist and Conservation Officers survey river or coastal areas, liaising with the engineers responsible for engineering projects, and do field work to protect and improve the status of rare creatures. Conservation issues include reed-bed and tree replanting, the protection of wild flowers and wading birds, and the preservation of wind and water mills.

Jobs with the NRA are advertised at local level, or in the *Guardian* and other publications. Graduate entrants to conservation posts, with degrees in biological or chemical science, are trained through work experience and supported in continuing education and studies for membership of professional institutes, including the Institution of Water and Environmental Management, the Royal Society of Chemistry and the Institute of Biology.

The environment protection section of the DoE employs a wide range of professional officers, many working in integrated teams with administrators. Together they develop and implement government policy towards protection of the environment including the supervision of a substantial research programme. The DoE represents the UK during discussion of most aspects of pollution control in international forums, notably in the European Community.

The Property Services Agency

The PSA is the largest design and construction organisation in the UK, with an annual budget of around £10 billion. It has a strong interest in the conservation of old buildings, as well as new, and has been involved in restoring the Palm House at Kew Gardens and advising on the new British Library in London.

The Department of Energy

The Department of Energy was set up in its present form in 1974. It is responsible for the development of national energy policy, including the management of current sources of energy and development of new sources, energy efficiency, representing the government in dealings with the nationalised energy industries (coal and electricity) and the United Kingdom Atomic Energy Authority, sponsoring the oil and nuclear power construction industries, overseeing government interest in the development of offshore oil and gas, and dealing with international energy affairs.

The Department employs about 1,150 people, of whom all but about 170 work at departmental headquarters in London. The work of the department is divided into different sectors, of which the following relate to conservation.

Energy Efficiency Office. This is the division with general responsibility for the government's energy efficiency policy. Activities include an educational programme aimed at furthering conservation, the promotion of efficiency measures, research into and development of new measures, and technical advice to industry, commerce, the construction industry, local authorities and other large energy consumers.

Energy and international policy. This division is responsible for the government's energy policy and for UK policy on international energy affairs. Activities include the environmental aspects of energy policy, relations with the European Community and other countries on energy policy, and energy pricing.

Energy Technology Support Unit. This sector consists of 40 professional staff of scientists and technical advisers who operate research, development and demonstration programmes dedicated to energy efficiency.

Information. This is the division responsible for the Department's public relations, press liaison, publicity and publications.

Other specialist divisions are assigned to work on nuclear energy, coal, electricity, gas, oil, and petroleum policy. They manage government input into these areas, review supply and demand, and carry out research into public safety.

Overseas Development Administration (ODA)

The ODA is part of the Foreign and Commonwealth Office, and formulates and carries out British development aid policy, much of which is directed at Third World countries.

Aid is in the form of financial and technical co-operation to help developing countries with specific projects (such as building roads, setting up fisheries, forestry research or emergency relief) and with long-term programmes (such as forest management, soil conservation, or energy development). A large amount of aid goes into conservation and resource management projects.

The ODA runs specialist departments dealing with projects in rural development, natural resources, science, technology, health and population, education, and manpower. It finances a number of specialist research and advisory organisations, and has its own scientific unit, the Overseas Development Natural Resources Institute.

Natural Resources Institute. This is the ODA's in-house scientific unit at Chatham in Kent, comprising some 500 scientific and support staff.

The Institute's mandate is to promote sustainable development of renewable natural resources in the tropics.

The main areas of work are: resource assessment and farming systems; integrated pest management; and food science and crop utilisation. The Institute has purpose-built laboratories and a world-class library, and the work covers applied research, surveys, the transfer of technology and programmes of advice and consultancy. The staff collaborate in scientific projects at home and overseas, and most travel overseas on short- and long-term assignments, to around 60 countries across the developing world. The main disciplines are chemistry, biochemistry, land use, livestock nutrition, food technology, engineering and economics.

Other ODA activities include research into water management and conservation, re-afforestation and forest management, and the development of energy and fishery resources.

The Natural Environment Research Council (NERC)

The NERC is one of the government's five Research Councils, and was set up in 1965 to research physical and biological sciences relating to the natural environment. With more than 2,500 staff, it is one of the largest employers of school leavers and graduates with qualifications and an interest in the life sciences.

Like the other Research Councils, the NERC is autonomous and operates under a Royal Charter. It is funded by the Department of Education and Science and carries out its research through a series of specialist institutes, and gives grants and awards for research in universities and other institutes of higher education. It also advises government and industry on environmental matters.

The NERC has 21 institutes, of which the following are of particular interest to environmental scientists.

The Institute of Terrestrial Ecology studies the ecology of land ecosystems, including changes in land use, the effects of pollutants, and threats posed to endangered species. It employs 249 staff.

The Unit of Comparative Plant Ecology studies the interaction of plants with their environment, and the mechanisms controlling plant distribution and vegetation structure. It employs 15 staff.

The Institute of Virology and Environmental Microbiology studies viruses, virus diseases of insects, and their effects on other forms of life. It employs 68 staff.

The Institute of Freshwater Ecology studies the ecological characteristics of inland waters and the ways in which water is used. It employs 100 staff.

The Institute of Hydrology studies the phases of the hydrological cycle and the problems of managing and using water as a natural resource. It employs 161 staff.

The British Geological Survey undertakes geological research to provide information for the exploration and use of mineral, water and energy resources. It employs 795 staff.

The Institute of Oceanographic Sciences Deacon Laboratory studies the characteristics of the ocean, oceanic resources, and the structure and topography of the ocean bed. It employs 162 staff.

The Plymouth Marine Laboratory carries out inter-disciplinary research into estuarine, coastal, shelf and oceanic ecosystems. It employs 149 staff.

The Dunstaffnage Marine Laboratory undertakes a variety of marine surveys, including environmental impact assessments and environmental management and monitoring. It employs 47 staff.

The Proudman Oceanographic Laboratory studies the theory, observation and numerical modelling of sea levels worldwide, shelf and slope processes and circulation. It employs 80 staff.

The Sea Mammal Research Unit studies the role of seals and whales in marine ecosystems and the effects of management policies on their populations. It employs 16 staff.

The British Antarctic Survey undertakes year-round research in the Antarctic into topics such as terrestrial, freshwater and marine ecology, and helps to formulate policy on the rational management of the Antarctic environment. It employs 414 staff.

Recruitment of staff is handled either directly by the Council, or by the Civil Service Commission, depending upon the grade of the vacancy.

Scientific staff make up the bulk of NERC employees, and vacancies are normally advertised in the national press, scientific journals, and universities. The most junior level is the *assistant scientific officer*, who needs four GCSEs or equivalent, including English language, a science or maths subject, and not more than one of an artistic, commercial or domestic nature. Promotion is to *scientific officer*, for which post candidates need to be aged under 27, and have a degree in a science, maths or engineering subject, degree standard membership of a professional institution, or a Higher BTEC Certificate in a science, maths or engineering subject, or equivalent.

Candidates for the post of *higher scientific officer* should have the same qualifications as a scientific officer, in addition to which they need at least two years' postgraduate research or development experience (if they have a first or second class honours degree or equivalent), or else at least five years' appropriate experience. They should normally be aged under 30. The most senior scientific post is that of *senior scientific officer*, for which candidates should be aged between 25 and 32, should have a first or second class degree in a science, maths or engineering subject, or equivalent, and at least four years' postgraduate or other approved experience.

Administrative staff are employed to support the scientific staff with clerical and executive duties. Candidates for the post of *administrative officer* should have five GCSEs or equivalent, including English, and should be aged over 17½. For *executive officer* posts, candidates should have two A levels or equivalent, passed at the same sitting, in addition to three GCSEs, and should be aged over 17½.

The Ministry of Agriculture, Fisheries and Food (MAFF)

As its name implies, MAFF is responsible for carrying out government policy on agriculture, fisheries and food.

Within that wide spectrum MAFF's concerns include:

- monitoring animal health and welfare;
- implementing environmental protection schemes;
- maintaining public health standards in the manufacture, preparation and distribution of foods;
- conserving the marine and freshwater environments.

Given that more than 70 per cent of our total land surface is farmed, and that fishing is one of our principal offshore industries, the degree of MAFF's influence over the natural environment is considerable.

MAFF seeks, in drawing up policies for agriculture, to achieve a reasonable balance between the interests of agriculture, the social and economic needs of rural areas, the conservation of the countryside and the promotion of its enjoyment by the public. Conservation interests are consulted when new policies are being formulated and a number of measures of direct benefit to countryside conservation have been introduced in recent years, including the establishment of Environmentally Sensitive Areas. MAFF employs about 10,000 staff, of whom about half are specialists in scientific or technical areas. About a third of the staff work at the London headquarters, and the remainder are at MAFF's regional offices.

The Directorate of Fisheries Research, a department of MAFF based at Lowestoft, researches the commercial exploitation of fish and shellfish stocks, and marine and freshwater pollution. There are openings for marine ecologists, biologists and fish biologists, biochemists, geneticists, chemists, physicists and geologists, and experts in fish disease, population dynamics and operational research. Most of the openings for environmental scientists and planners are with ADAS.

ADAS

ADAS (Agricultural Development and Advisory Service), became an executive government agency in April 1992. The agency is owned by MAFF and the Welsh Office. ADAS has provided advice and consultancy to farmers and growers for nearly 50 years and since the mid-1980s consultancy has extended to all operating in land-based industries, especially those involved with food production, processing and retailing, land management, waste and energy. ADAS's 100,000 customers include a broad mix of farm-based enterprises, independent food processors, manufacturers, retailers, utilities, waste managers, mineral extractors and land developers.

ADAS consultants operate throughout England and Wales, carrying out initial enterprise appraisals, market research and strategic planning and advising on the business and technical aspects of production, quality control and distribution. Project teams provide practical solutions to key environmental problems affecting users and developers of land. ADAS is also directly involved in the input to policy formulation and provides guidance on issues relating to food hygiene, animal health and environmental management and planning.

For further information contact ADAS Headquarters, Oxford Spires Business Park, The Boulevard, Kidlington, Oxford OX5 1NZ.

The Forestry Commission

Since it was founded in 1919, the Forestry Commission has been the government body responsible for the nation's forests. It owns just under half of Britain's 2 million hectares of productive forest, and is responsible for promoting the interests of forestry, for the efficient production of wood for industry, for the development of afforestation, and for the welfare and conservation of wildlife in the forests. Many areas within the Commission's woodlands have been designated Sites of Special Scientific Interest and National Nature Reserves, and a few have been set up as Forest Nature Reserves. The overall policy is to balance the needs of wildlife with those of the timber industry.

Openings within the Forestry Commission are different from those in the private forestry sector (see Chapter 4). The Headquarters of the Forestry Commission are in Edinburgh. The Commission reports directly to three Forestry Ministers - the Secretary of State for Scotland, the Minister of Agriculture, Fisheries and Food in England and the Secretary of State for Wales. To carry out the government's aims efficiently and effectively, the Forestry Commission has three distinct sub-divisions - the Forestry Authority, Forest Enterprise and the Policy and Resources Group.

- *The Forestry Authority* administers the woodland grant scheme, carries out forestry research and monitors standards in all forestry, including the Forest Enterprise.
- *Forest Enterprise* has sole responsibility for the management of the Commission's forest estate. The government now intends to replace Forest Enterprise with a trading body set up as a Next Step Agency from April 1995.
- *Policy and Resources Group* is responsible for the parliamentary and policy aspects of the Commission's Departmental duties and provides services, such as personnel management and business systems, to The Forestry Authority and Forest Enterprise.

The *Forest Enterprise* has five regional offices as follows: York, Bristol, Aberystwyth, Inverness, Dumfries.

The *Forestry Authority* has three National Offices as follows: Cambridge (England), Aberystwyth (Wales), and Glasgow (Scotland).

Each Enterprise Region is divided into Forest Districts and each Authority National Area is divided into Conservancies.

A Director of Research Division, stationed at HQ, Edinburgh is responsible for such specialised fields as research, statistics and technical publications.

Forestry jobs are at two levels: forest worker and forest officer III.

There are occasional vacancies for wildlife rangers who look after the forest environment.

Forest worker. Anyone who is physically fit and interested in manual work outdoors can apply for work as a forest worker. The work is mainly manual or operating machines in the forest with a wide range of work including fencing, planting, draining, weeding, pruning, timber harvesting and nursery work. Training is given to enable forest workers to qualify as forest craftsmen and for further advancement to ranger and foreman or to wildlife ranger. Jobs are advertised locally.

Forest officer III. A recognised qualification in forestry is essential and candidates must either have:

a. a BTEC or SCOTVEC Higher National Diploma in Forestry; *or*
b. A BTEC National Diploma in Forestry; *or*
c. a Degree in Forestry; *or*
d. a SCOTVEC National Certificate embracing all supervising and management level modules appropriate to Forestry; *or*
e. a City and Guilds of London Institute (CGLI) Phase IV Certificate in Forestry; *or*
f. be a final year student who expects to obtain one of these qualifications in 1995.

The ability to drive a car is essential.

Officers in this grade are technical forest managers and are responsible for planning annual work programmes, supervising and training Forest Workers, estimating costs, setting piecework rates, and measuring and controlling work programmes. They are responsible for protection of the forest, for safe working practices, for relations with neighbouring land owners and with organisations and individuals who wish to use the forest for sport or recreation. They are also employed on specialist duties, such as work study, wildlife conservation, research and training.

Wildlife ranger. Formal qualifications are not required but relevant previous experience is essential. All wildlife rangers are thoroughly trained.

The work entails protecting and conserving the forest environment, controlling pests and protecting forest wildlife, as well as creating and maintaining suitable habitats. Rangers also guide and assist the public who visit the forests and enforce the by-laws. A 24-hour commitment to the job is necessary.

The Forestry Commission also has administrative assistants, administrative officers and executive officers in its headquarters and offices, and there are openings for scientists, mechanical and civil engineers, land agents and clerks of works.

Career prospects. Forestry is primarily concerned with growing trees for commercial exploitation, and as such requires an understanding of the needs and structure of the forest environment. Without conservation, exploitation could not be sustained. Competition for entry to forestry is high at all levels.

Case Study
Ben is a forest worker with the Forestry Commission. He left school when he was 16 and did a variety of odd jobs before applying for his current job, which he has been doing for just over two years. He was attracted to it originally because it meant working out of doors, but now he sees it as a career.

> I'm still on the first rung of the ladder, but I am hoping to stay with the Commission and work my way up. At least this way you pick up working experience as you go along. It's mostly very hard work and it can get a bit boring at times, but there is something exciting about working in forests. I'm part of a team working under a foreman, and we do a whole variety of jobs. Just before the planting season you could be doing fencing and maintenance work, then you get down to planting transplants during the early spring. The summer is taken up with weeding and clearing the new plantations, and then the thinning begins in the autumn. In between I've been sent off on short courses on how to use and maintain machinery and forest management. I would like to keep on working out of doors all my life, so I want to avoid being promoted to a desk job within the Commission. I've got my eyes on a job as a warden eventually.

Environmental Planning for Local Authorities

An increasing population and growing pressures on land have heightened the need for effective planning systems in recent decades. Countryside planning is of particular interest to anyone wanting a career in resource conservation. The conflicting demands of agriculture, amenity, recreation and conservation have to be met, and the countryside planner has much influence over the overall outcome.

The structure of planning in the United Kingdom was laid out in the 1947 Planning Act, significantly revised in 1971, which required planning authorities to prepare development plans for the allocation of land for roads, housing, industry, open space, agriculture, conservation and other uses. Private and public development was also controlled to fit in with the overall scheme for the country, with the result that the UK now has a comprehensive planning structure. It has not always been ideal, but it provides the rule against which people with conflicting demands can measure the allocation of land.

About 60 per cent of the openings for planners are in local government. The Department of the Environment is the central government body which supervises planning. It publishes the policies of the government, issues guidance on practices and procedures, commissions research, and provides local planners with information. At the regional level, county councils (regional councils in Scotland) draw up broad strategic plans for the development of each county, based on careful surveys of the county's economic and physical resources.

At local level, district councils prepare local plans, which go into more detail than the county structure plan, and also oversee development control by dealing with applications to develop or change the use of land. This is the area that directly affects all our lives, and most directly affects the quality of the environment. Plans and changes at every level must in turn be subject to public inquiries, particularly if there are objections.

Outside central and local government there are opportunities for planners in statutory bodies such as the Countryside Commission, water authorities, national park planning authorities and development corporations. There has recently been a growth in the number of independent professional planning consultants. Private consultancies are often commissioned to work on special schemes for local government, but otherwise operate mostly with institutions such as universities, or work on private sector development schemes. There is expected to be an increasing amount of work abroad, especially in the EU from 1992.

The job of the planner is to help in preparing development plans, which involves carrying out all the necessary research, survey and consultation work in the field and working with experts in related areas; or it might be to deal with applications for planning permission. Planners work as part of a team, and might be asked to work on a variety of jobs, from working out population trends to improving the environment by planting trees or reclaiming derelict land. The plan could be a general one for an entire county, a specific local one for a country park or a coastal development, or an environmental scheme within a small area. Planners have to consult the public, politicians, officials, industrialists, financiers and technical experts concerned with conservation and development.

Qualities Required

Planners must be able to work together in teams and with experts and advisers in different fields. They must have a great interest in the environment, and understand the needs of the community, and must be able to communicate their ideas and plans to all the people with whom they come in contact. They should be broad-minded, logical, willing to listen to and take advice from others, able to back up their ideas with careful research, have the self-assurance to back up their decisions, and be able to cope with the many frustrations that come with the job.

Qualifications

Membership of the Royal Town Planning Institute (RTPI) is the recognised professional qualification for town planners in the UK, and this is accepted in many other countries as well. Candidates have to satisfy the Institute's academic and practical training requirements, and must have had two years' practical experience. About 95 per cent of entrants enter planning via a course at a school of planning or an accredited university, and can become student members of the Institute. There are undergraduate courses for school leavers, or postgraduate courses for graduates with relevant first degrees, usually in architecture, surveying, engineering, geography, economics or sociology. The alternative to a college course is to study at home on the RTPI accredited distance learning course.

Openings and Prospects

Job prospects for qualified planners are reasonably healthy at present but do fluctuate with the economic climate and the level of local and central government spending. The RTPI currently has over 15,000 members and student members. The work of planners is changing, with many people now involved in areas such as economic development, conservation and tourism, as well as the more traditional areas of development control and local planning. It can also be useful to back up a qualification in planning with skills in related fields such as architecture and surveying.

Chapter 6

Working for Statutory Bodies

Introduction

Much of the responsibility for managing wildlife and natural resources in the UK lies with statutory bodies - that is, bodies set up under law. They lie somewhere between the independent and government sectors, being funded largely by the government but not actually government bodies.

The main bodies entrusted with conservation and the environment are the three new country conservation bodies which have taken over from the former Nature Conservancy Council that covered the whole of the UK. They are English Nature, the Nature Conservancy Council for England, Scottish Natural Heritage (formed by a merging of the Nature Conservancy Council for Scotland and the Countryside Commission for Scotland), and the Countryside Council for Wales; plus the Countryside Commission, 11 national park authorities, and the Forestry Commission which is a full government department (see Chapter 5). Between them they offer a range of openings for administrators, planners and scientists.

English Nature

English Nature advises the government on nature conservation in England, and promotes, directly and through others, the conservation of England's wildlife and natural features within the wider setting of the UK and its international responsibilities. It is based at Peterborough, at the headquarters of the former Nature Conservancy Council, and its work continues to be to select, establish and manage national nature reserves, and to identify and notify the Sites of Special Scientific Interest. There are currently over 150 nature reserves and 3,800 SSSIs. The Council provides advice and information about nature conservation and support, and conducts relevant research.

English Nature works with the Scottish Natural Heritage and the Countryside Council for Wales on UK and international conservation issues through the Joint Nature Conservation Committee. (The Department of the Environment for Northern Ireland has parallel responsibilities.)

Although it is not a government department, English Nature offers the same rates of pay, conditions of service and recruitment regulations as the Civil Service, and demands certain minimum qualifications or experience.

Openings

Scientific staff. Openings here are mainly for graduates, and the usual point of entry is the *assistant conservation officer (ACO)*. The ACO is the Council's public face, and is responsible to the *conservation officer* for a variety of duties. ACOs spend most of their time identifying habitats with high nature conservation interest. This involves surveying and assessing scientific sites, working with the owners on the conservation of statutory sites, working with local planning authorities on any developments that affect wildlife and nature, consulting voluntary conservation bodies, and conservation education.

A degree in a relevant subject, or degree standard membership of a professional institute or HNC or HND, is usually the minimum requirement for ACOs.

Occasionally vacancies arise for conservation officers. Qualification requirements are the same as for ACO, with additionally at least five years' relevant experience. If applicants possess a 1st or 2nd class degree then only two years' postgraduate research (usually MSc or PhD) is needed. Previous experience is usually the deciding factor, so any voluntary work undertaken with clubs, societies, charities or natural history groups would be useful.

Vacancies are advertised in the *Guardian* and *New Scientist*, and competition is usually stiff.

Reserve staff. If competition for *scientific staff* vacancies is stiff, then competition for posts as *site managers* for the reserves is fierce. There are more than 150 National Nature Reserves, and the Council needs site managers to oversee their management, which includes everything from supervising voluntary wardens and estate staff to patrolling the reserve, carrying out maintenance, advising and guiding visitors, conserving wildlife, carrying out scientific recording and monitoring, maintaining records and reports, servicing scientific research work, maintaining nature trails, and liaising with neighbouring landowners.

Applicants must be at least 26 years old, with five years' paid experience in the field of nature conservation, and proven ability in the field of natural history. They must also hold a full current driving licence. Formal educational qualifications are not compulsory. Jobs are advertised in *New Scientist.*

Estate Workers provide the manual labour for land management of reserves. They may be required to work alone or in teams under the supervision of site managers. Vacancies do not arise often. Jobs are advertised in the local press and Jobcentres and recruitment is carried out by the local team office.

Executive and clerical staff. Executive and clerical openings follow normal Civil Service procedures, but recruitment is managed by EN. Vacancies for executive grades are usually filled internally. Recruitment for clerical posts is carried out by the local team office.

Career Prospects

English Nature is a relatively small organisation which attracts a good deal of interest. Job opportunities tend to arise fairly infrequently and competition is therefore quite high.

Vacancies, when they occur, are normally advertised in either the local press, Jobcentres, the *New Scientist*, *Guardian* or specialist publications where appropriate.

Scottish Natural Heritage

Scottish Natural Heritage is a merging of the Nature Conservancy Council for Scotland with the Countryside Commission for Scotland, on April 1, 1992. The responsibility of Scottish Natural Heritage includes 'the flora and fauna of Scotland, its geological and physiographical features, its natural beauty and amenity'. The work is similar to the work of English Nature, including managing National Nature Reserves and Marine Nature Reserves, selecting SSSIs, conservation of protected plant and animal species, research, and raising awareness about nature conservation. SNH projects include the conservation of an endangered species of plant or animal, special habitats, helping schools create wildlife gardens in their grounds and the regeneration of native woodland.

Between them the two groups forming SNH have a staff of over 460; their skills cover ecology, geography, geology, biology, communications, education, landscape architecture, planning, land use management, archaeology, graphics and cartographics, and many other fields.

The main types of employment are similar to English Nature and the old Nature Conservancy Council, with Assistant Regional Officers, Scientist Officers, Earth Science Division, Field Survey Units and Specialist Development Officers, Reserve Wardens and Estate Workers, plus executive and administrative staff.

Countryside Council for Wales

The Countryside Council for Wales (CCW) is the organisation that deals with countryside matters in Wales on behalf of the government. It provides advice on countryside and maritime matters to ministers and government departments as well as enabling others, including local authorities, voluntary organisations and interested individuals, to pursue countryside management projects through grant aid.

CCW advises the government and local authorities on matters that affect SSSIs, National Parks, Areas of Outstanding Natural Beauty (AONB), and on relevant Environmental Impact Assessments. The council also advises and comments in response to consultation on development plans and local planning applications, and takes part in public consultation exercises at national and local levels. CCW would also respond to EC Directives that affect its development in Wales through the Joint Nature Conservation Committee, in conjunction with the related bodies in England and Scotland. The advice given by the Council is based on the experience of conservation bodies, ecologists and countryside managers, gained since the 1949 Countryside Act.

Advice is offered to owners and occupiers of rural land, voluntary organisations committed to the care of the countryside and waters.

The Council employs over 200 staff throughout Wales, including scientists, rangers and wardens, and also uses expertise provided by specialist organisations.

Wales has three National Parks, more than 50 National Nature Reserves, five AONBs and about 800 SSSIs.

CCW is responsible for conserving the natural features and wildlife of Wales and the intrinsic quality of the landscape. It also promotes opportunity for access to the countryside.

The Countryside Commission

The Countryside Commission replaced the National Parks Commission in 1968, and was charged with responsibility for the conservation and enhancement of the countryside and promoting access and the provision of recreational facilities. Then part of the Civil Service, it now has greater independence and manages its own recruitment for posts. There is one Commission for England. Those for Wales and Scotland are merged into new country conservancy bodies.

The Commission's responsibilities include designating and supporting (with grants and advice) national parks, areas of outstanding natural beauty, national trails and bridleways and the Community Forests Scheme. It also has a research and experimental programme and provides publicity and information about the work it is doing. On a national level it gives advice to government on countryside policy generally, and on a local level to conservation bodies, landowners and local authorities.

Openings

The Commission employs about 350 staff, half of whom are employed in the head office at Cheltenham and the other half in the network of eight regional offices. There are three divisions:

1. **Resources**, which is responsible for the provision of support services including programme planning, finance, personnel and office services.

2. **Operations**, which consists of head office staff and seven regional offices and the office for Wales employing between four and seven staff in each office, to promote the Commission's policies regionally, work with local authorities and individuals, and manage grants for countryside work.

3. **Policy**, which consists of five branches at head office which advise on and develop policy in their particular subject areas.

Land Use Branch. Landscape conservation, monitoring of land-use trends, including agricultural policies, forestry and woodland management (multi-purpose forestry), water and land drainage. The branch employs suitably qualified life scientists, planners, landscape architects and economists.

Recreation and Access Branch. Country parks, national trails and bridleways, rights of way and access, commons, 'enjoying the countryside', management of countryside for people. It also advises on information, countryside staff training, and visitor services.

National Parks and Planning Branch. National parks, areas of outstanding natural beauty and heritage coasts. It also covers development control and planning issues.

Environmental Protection Branch. Community action, voluntary bodies, voluntary and paid work in the countryside. The branch provides policy direction on voluntary body issues and employment and training in the countryside.

Public Affairs Branch. Press relations, publications, general advice to the public and the library.

Career Prospects

Because of its relatively small size, the Countryside Commission offers few openings, and because it owns no land it does not employ part-time staff in the same way as the other two statutory bodies. For the time being it offers planners, scientists, and national and regional administrators the chance to be involved closely with national conservation activities. Recruitment is handled nationally.

National Parks

Although the Countryside Commission is responsible for the general management and welfare of national parks, each is controlled by its own board or committee. There are 11 national parks in England and Wales (Scotland has none), of which two (the Peak and the Lake District) have their own park boards. Dartmoor, Northumberland, the Pembrokeshire Coast, Snowdonia, Brecon Beacons, Exmoor, the North York Moors and the Yorkshire Dales are run by county committees and the Broads Authority, which has similar status to a National Park Authority, is responsible for the Broads.

Openings

Each board is autonomous and is responsible for planning, buying land, administration, recruiting staff, and determining salary grades of employees. So it is difficult to generalise about opportunities, although all the parks tend to employ common grades of staff to carry out broadly similar duties.

The annual budget of each board varies considerably. This is reflected in the number of staff employed: the Peak is the largest employer, with over 150 full-time staff and 200 part-time and seasonal staff, and

Northumberland the smallest, with approximately 30 in four main grades of service.

Management, administration and planning. This area consists largely of senior or professional staff charged with the management of the parks, who are responsible for liaising with the owners of public and private land that falls within the borders of each park. They resolve the different demands of development and conservation, manage budgets, buy new land, and plan and provide guidelines for the management of the parks. There are openings for professional planners, surveyors, park and recreation administrators, and other executive and clerical staff. About 40 per cent of national park staff are employed in this sector.

Information. Because national parks exist for the enjoyment of the thousands of visitors who visit them annually, the provision of visitor centres and information facilities is a major function. Staff with a background in information and education are employed to run the centres and to work directly with the public. Duties include setting up displays, publishing leaflets and guides, working as sales assistants, running educational facilities, and providing general guidance.

Wardening. The daily management and upkeep of the parks is in the hands of wardens and rangers, who are responsible for guiding visitors, carrying out conservation policy, protecting wildlife, supervising management work, helping with information and interpretation, and keeping in touch with local landowners. The qualifications necessary are similar to those for Conservancy Council wardens: a high degree of experience, relevant skills and abilities and a deep interest in natural history and conservation.

Project and estate management. Carrying out estate projects is the job of a combined team of professional estate managers and manual workers. They maintain recreational and visitor facilities, look after roads, paths and nature trails, carry out general maintenance of the parks, operate machinery, and undertake other tasks.

The parks have openings for scientists, planners, administrators, and executive, clerical and manual staff, and some employ ecologists, naturalists, foresters, soil scientists and other conservation specialists.

They also employ a substantial number of part-time and seasonal staff, mostly during the summer season when the number of visitors increases sharply. Many of these staff are students working in their vacations, school leavers, and members of government job-creation schemes. They are employed as information assistants, weekend wardens, seasonal rangers, student researchers, car-park attendants and services staff. Vacancies are normally open to existing staff before being advertised publicly.

Career Prospects

The level of opportunity varies with the economic climate and the budget policies of local authorities. Despite the fact that salary levels are generally below par, competition for vacancies is always keen, so candidates need a high degree of ability and experience. Working in one of the national parks means being involved in a central area of national conservation activity.

Case Study

George has been working as a national nature reserve warden for the past five years. He was born and brought up only five miles from the reserve, so he knows the area intimately.

> It was the reserve that first got me interested in natural history and conservation, when my uncle used to bring me along to bird-watch when I was at school. He knew everyone who worked here, and it was only a matter of time before I was working here at weekends and helping out with the information centre. After leaving school I took an environmental studies course at college, but even though I saw it through I am not sure it helped to prepare me for the practicalities of working on a reserve. You learn mostly through experience, and from being taught by the older hands. My experience as a voluntary warden was what really got me this job – I heard that there were 250 applicants.
>
> I love the work. The pay is poor and we sometimes feel a bit forgotten by head office, but I'm really my own man and it is up to us to make sure that everything is run properly, that good facilities for visitors are maintained, and that people get the most out of coming here. You also feel a strong responsibility for the welfare of the wildlife. The job involves a lot of research and surveying, as well as routine reserve maintenance. It's hard work, with long hours out of doors in rain and shine, but time doesn't matter because it's more a way of life than a job.

Common Sectors

Chapter 7
Education and Research

Introduction

Pioneer conservationists tended to 'patch up' problems wherever they surfaced rather than look for the underlying causes. Part of the problem lay in the fact that they knew (and we still know) little about the structure of the natural environment and about the threats it faced and faces. The knowledge we now have has been accumulated in a short space of time, and we have probably learned as much in the past 20 years as we learned in the preceding two centuries. There has been an increased emphasis given to both environmental education and research in conservation in recent years.

Environmental Education

Education is now widely regarded as being one of the most important environmental priorities, but much of the initiative for its development still rests with the voluntary sector. Environmental studies is still a rare option in British secondary schools, and ecology is barely taught at all. Geography, biology, botany and zoology are, on the other hand, widely taught, and occasionally give rise to more specialised fields of environmental study.

The options are much wider in higher education, where conservation, ecology and environmental studies are more frequently offered alongside the life sciences and professional training in planning, forestry, estate management and related topics. Outside the education establishment, a number of voluntary bodies are developing increasingly thoughtful education programmes.

Teaching in Schools

The teaching profession is a career option that offers a number of openings for environmentalists. The number of schools offering environmental studies and science as GCSE and A level options is growing, and a number of organisations (notably the Council for Environmental

Education and the National Association for Environmental Education) both promote environmental teaching in schools and offer students and teachers guidance and assistance.

Teachers who run classes in almost any subject can often touch on environmental matters within the normal curriculum. Alternatively, a teacher who expresses an interest in teaching environmental studies may well get a sympathetic hearing from the school authorities, so it may be up to the teacher to take the initiative.

Training is via a course leading to a degree and qualified teacher status (QTS) or PGCE. Entrants to teaching are normally graduates and need a minimum of five GCSEs, including English language and mathematics, and an additional two A levels would be useful, especially in environmental studies or science. Entrants to universities generally need five GCSEs and two A levels, and following graduation they would study for a one-year Postgraduate Certificate of Education (PGCE).

Teaching in Higher Education

As well as offering courses in every branch of life science, many universities and colleges run professional courses in resource management, planning and conservation, and specialist courses in subjects such as plant ecology, pollution control, the management of genetic resources and alternative sources of energy.

Nearly all these institutions are currently suffering from spending cuts, so there is little new building. New departments are not being opened, and there are few vacancies. Jobs are therefore at a premium and even a high level of experience and academic achievement does not guarantee employment.

For those who are able to find jobs, entry is almost always through postgraduate research and study. Many university tutors and lecturers combine teaching with research and administration, so research interests and abilities are essential. Former polytechnics, now with university status, are not as much involved in environmental research at present.

Most institutions of further education are run by local authorities. They may either specialise in vocational training or offer a more general range of education. They vary considerably in size and in the scope of their courses. Part-time and evening classes in specialist minority subjects offer scope for the environmental teacher.

The minimum qualifications for university teaching vary. Someone with a firm background in research and a high level of academic achievement is likely to be favoured, but this is by no means certain. The demands are similar for polytechnic teaching posts but working experience is also useful, especially if it is in a field related to the teaching post. Someone wanting to teach environmental studies, for example, would probably need a few years' experience in resource planning and management or conservation.

Conservation Education

Many environmental and conservation groups, both voluntary and statutory, run national and local education programmes that employ qualified teachers and youth group leaders. Some programmes are designed largely to promote the aims of individual bodies by involving young people in conservation and the appreciation of natural history, while others have more general aims. Many young people take part by virtue of their membership of the body, while others take part in projects run by different bodies in their areas of interest.

So, for example, a charity might run a schools lecture programme, a natural history society might take parties of schoolchildren around a local nature reserve, an ornithological group may involve children in bird-watching and surveying, and a practical conservation group may run projects for young people at weekends and in school holidays.

Most of the larger conservation bodies have been running youth membership schemes for many years. They were mainly limited to subscription schemes, with children paying their dues and receiving magazines and other material in return. The current trend is towards programmes that involve children actively in conservation, coupled with efforts to encourage schools to include environmental studies in their curricula.

Voluntary organisations employ education officers and youth group leaders to run these programmes. Teaching experience is desirable, but by no means essential; a proven ability to run teaching projects and an interest in working with children and schools are the minimum requirements. Much greater emphasis is being given to carefully structured education programmes.

Other Fields

There are also openings in adult and continuing education, with local education authorities and private field centres, and with a few correspondence colleges. The adult and extra-mural education field offers short part-time courses that cater for all interests. Teaching is conducted mostly by part-time staff, and most posts go to people with part-time teaching experience. There is considerable potential for anyone interested in teaching specialist environmental courses on a part-time basis.

The Field Studies Council

The Field Studies Council, founded in 1943, is an independent charity that runs 10 field centres in England and Wales. These hold short residential courses in environmental studies, ecology, biology, geography and geology.

The Council runs courses for teachers to help them teach environmental studies in their own schools. There are courses in project and field work, analysis techniques, teaching ecology, geography and GCSE field work. The Council also employs graduates in full-time teaching posts and non-graduates in temporary summer work. Competition for the few full-time posts is very keen. Each Centre is run by a warden, assisted by two, three or four tutors, who may be helped by research assistants and/or conservation officers, plus administrative and domestic staff. Minimum qualifications for an academic post are an honours degree in an appropriate subject, and a teaching qualification is an advantage. The courses are run between mid February and early November for sixth-form, GCSE and younger pupils and for teachers, university students, professional people and adult amateur naturalists. This leaves the winter months largely free for research and course preparation.

Case Studies

Anne runs a youth education programme in Wales for a national environmental group.

> I have wanted to work in conservation as long as I can remember. From about the age of five I was a member of just about every junior wildlife watchers group that was going, and I took part in a lot of projects while I was at school. I took A levels in zoology and botany and then read zoology at university. It was while I was there that I began to realise that getting my ideal job in conservation wouldn't be as easy as I had thought. Graduates from two or three years before my year all seemed to be working in anything but zoology.
>
> About six months before I graduated I started applying for likely jobs, but had had no luck by the time I graduated. I was determined not to go on the dole, so I enrolled for a postgraduate certificate of education. I applied for jobs continuously for a year - even ones that were only vaguely related to conservation. Altogether I went through 56 applications from which I got 14 interviews but no jobs. My 57th application was for a secretarial job in the membership department of a conservation group, and I was accepted. Just by luck, the offer of the job I'm now in came up. It's working for the same organisation, and I think it was because they knew where my real interests lay that I was given preference over outside applicants.
>
> The job is everything I want for the time being. I was given virtual *carte blanche* to set the scheme up, and I'm now worked off my feet arranging field trips, writing and designing information packs, commissioning posters, producing audio-visual programmes, and starting up school groups. I'm the first to realise how lucky I have been to get this job - it was certainly worth the wait.

Bernard is a postgraduate student who supports himself by giving evening classes at a college of further education.

> I was fortunate enough to have a superb biology teacher at school who gave me my interest in natural history. She encouraged me to go on and read environmental studies at college with the idea of going on to teach. I got my first teaching post in an unusual place: holding classes on conservation for commuters on the train to college in my final year. I had to commute 90 minutes each way, and groups of commuters were getting together to hold classes to relieve the boredom. One day a college lecturer joined the group; we got talking, and he recommended that I approach a few colleges of further education and ask about part-time posts. I applied to three or four colleges and was lucky enough to get my present post, which I am now using to support me while I read for a master's degree, something which I didn't think I would be able to do for years.
>
> The job is mostly quite good fun, because I have a rapport with the students, who are nearly all older than me and don't mind getting into a few arguments and debates. Occasionally I have to take small classes of passive students, and I also get a bit frustrated by the irregularity of attendance. On the whole, though, I'm getting good teaching experience.

Research Opportunities

Research is a process of inquiry, scientific study and critical investigation - the collection, analysis and interpretation of facts. Because knowledge is not an absolute quantity - we can never know everything there is to know about something - there will always be scope for research. As far as our understanding of the natural environment is concerned, we are only just beginning to collect the facts. The active study of ecology is less than 100 years old, and we still do not fully understand the mechanisms of the biosphere. There are almost limitless opportunities to pioneer new avenues of study. The only constraints are financial. Perhaps because research is less obviously productive than teaching or applied fieldwork, it often suffers in the wake of spending cuts. And yet without research, knowledge suffers.

There are research opportunities in four main areas.

The Private Sector

Voluntary bodies, non-governmental organisations, and research institutes are responsible for the bulk of existing research programmes.

Charities either carry out their own research or support independent research with grants. Most of this support goes to universities, research institutes, and individuals with a reputation in research. However, almost anyone who can draw up a useful project proposal, give some proof of their own suitability, enlist the support of referees, and convince a charity that the project deserves support will be eligible for assistance.

There are several private research institutes carrying out research into population issues, Third World development, resource management, and the life sciences. They employ staff in a variety of capacities: visiting fellows, senior researchers, scientific officers and research assistants. Normal requirements are a first degree or equivalent qualification, combined with research experience and proven aptitude. The institutes work extensively with universities and the government, and maintain high standards of academic research and achievement.

Academic Research

The amount of research undertaken by universities and other centres of higher education that offer courses in environmental conservation and management is comparable to that of the private sector. Faculties, departments and affiliated research units generally devote more time to research and investigation than to teaching, and make considerable contributions to both general environmental knowledge and to specialised fields of study. As with teaching, finding a job in academic research demands a high level of experience and proven research ability, generally coupled with the possession of a higher degree and published work. Every aspect of conservation, resource management and the life sciences is scrutinised and studied.

Government Research

The government employs about 500 research officers to study the impact and implications of government policy and to help determine future needs. Many work for the Resource and Planning Research Group (RPRG), which studies policy on resource management and the environment, and some work on planning policy in the Department of the Environment.

In the Ministry of Agriculture, Fisheries and Food most scientific research work is carried on at the ADAS Central Sciences Laboratories at Slough, Tolworth (where a group studies the effect of pesticides on wildlife), Worplesdon and Harpenden; research into fish cultivation and the effects of pollution on the aquatic environment is carried out in the Directorate of Fisheries Research laboratories at Lowestoft, Burnham-on-Crouch (non-radioactive marine pollution), Conwy and Weymouth. There is also some research into pollution at MAFF's Food Safety Directorate Torry Research Station at Aberdeen. Research officers are also employed in the Agricultural Scientific Services and Fisheries Research Services of the Scottish Office.

The Laboratory of the Government Chemist, the government's focal point for analytical science, has a programme supporting the protection of the environment, scientists analyse foodstuffs, agricultural materials, and toxic metals and other hazardous materials of environmental interest.

Those working for the Overseas Development Administration are at the Natural Resources Institute, providing information on the economics of plant and animal resources in developing countries. In the Ministry of Agriculture, Fisheries and Food, most researchers work for the Agricultural Development and Advisory Service on land use studies. Research officers are also employed in equivalent capacities in the Welsh Office and the Central Research Unit of the Scottish Office.

Most government research officers are qualified in geography, agricultural economics, economics, economic geography, or an equivalent area.

The Industrial Sector

People in commerce and industry are gradually taking more interest in the environmental impact of their activities, and they are supporting more research, either internally or by providing grants to research institutes. The number of scientists and planners employed in this work has grown over the past 20 years, but the field is still restricted. Few companies employ more than a handful of specialists, who will require a high level of experience and proven research ability. One of the largest employers of research scientists is ICI. The laboratories of ICI Agrochemicals Jealott's Hill Research Station, employing over 120, and Group Environmental Laboratory at Brixham are among several in the company studying environmental toxicology. Areas of research include discharges to rivers, lakes and the sea, problems of air pollution, the disposal of waste to land, and contamination of soil and groundwater. As with all areas of investigation into the natural environment, the industrial research sector is likely to expand.

Chapter 8

Opportunities for Scientists

Introduction

Science is the backbone of conservation, because without an understanding of the structure of the natural environment there can be no really coherent attempts to protect and maintain it; conservation has even been described as applied ecology. On that basis, it would seem logical that a training in life sciences would be the best of all possible backgrounds for a conservationist, and that it would almost guarantee a job. In fact, this is not so.

Amateur naturalists in the nineteenth century were largely responsible for the emergence of the notion of conserving habitat, but it was only when the work of biologists, botanists and zoologists revealed the interdependence of species and the complexity of the natural world that conservationists began to realise the importance of science.

Environmental management requires co-operation between people from several disciplines, social scientists and economists as well as biological and earth scientists. There are now a few degree courses combining these disciplines, as well as a larger number of postgraduate qualifications. For those wishing to enter the practical types of conservation work, rather than research and theoretical development, these broad degrees are a better preparation than those confined to one or two disciplines.

Biology

Biology is the study of life and its structure: the form, functions, anatomy, behaviour, origins, and distribution of the earth's plant and animal species. Because it cannot be studied in isolation from other life sciences, it demands broad scientific training. Biologists study the form and structure of living organisms, the characteristics of ecosystems, the effects of environmental change, and a host of related subjects.

Biologists tend to work in health services, local authorities, water companies, the National Rivers Authority, landscape design, environmental consultancy, agricultural and industrial research, food science, and medicine. Most large engineering and oil companies also employ biological ecologists as well as geologists. The Game Conservancy trains and employs biologists for management of game reserves, and biology is a useful discipline for land agents on large estates.

Botany

Botany is the science and study of the structure, form, function, distribution and classification of plants. It is one of the oldest life sciences, studied for centuries by amateur naturalists attracted by the proximity, beauty and abundance of plant life. Despite this heritage, it is only now coming into its own as an environmental science. Until about 10 to 15 years ago, plants came a very poor second to animals in the priorities of conservation. The realisation that no animal species can be protected in isolation from its natural habitat, the foundation of which is nearly always plant life, and that perhaps as many as 25,000 plant species are threatened with extinction, has finally brought botany and plant ecology to the forefront of conservation.

Zoology

Zoology is the study and science of the structure, functions, habits, distribution and classification of animal life. The zoologist studies the anatomy and physiology of species, their feeding, territorial and reproductive habits, population trends, the characteristics of their natural habitats, and the ways in which they interact with their environment.

There are some openings for zoologists in conservation.

Ecology

Ecology is the study of the relationship between plant and animal life and its natural environment. Because the web of life is so complex, relationships are many and varied. Ecologists study the impact of species on their habitat, population trends, the influence of climate on life, the food chain, reproduction, the structure of ecosystems, and the impact of human activity on nature.

Many life science degrees have a substantial ecological component, and several are primarily ecological.

Marine and Freshwater Biology

Marine and freshwater environments are the subject of a considerable proportion of biological research activity, so much so that marine and freshwater biology has become almost a science in its own right. The field covers the study of animal and plant life in rivers, lakes and the open sea, the management of water and fishery resources, the effects of pollution, and the characteristics of marine and freshwater ecosystems. Because it is a specialised field, the number of openings is restricted.

Oceanography

Oceanography is the study of the oceans: the pattern and fluctuations of currents, the distribution of marine animals and plants, water temperature and salinity (salt level), and the geophysical structure of the ocean floor.

Until recently the oceans have been thought of as a dumping ground, but more attention is now being paid to oceanic resources such as fisheries, offshore oil and minerals, with the result that oceanography, like marine biology, is assuming a more important role in the management of the marine environment.

Most oceanographers tend to be qualified in biology, zoology, physics or geology. There are very few courses in oceanography.

Biochemistry

Biochemistry is the study of the chemical and physical processes involved in the structure of animal and plant life, with emphasis on the form and functions of organisms. It makes a valuable contribution to the understanding of organic life, and may determine the basis of local ecosystem conservation. However, most biochemists opt for careers in research, education, medicine, and the agrochemical industry.

Soil Science

Soil is one of the most valuable of all natural resources. It supports not only agriculture but almost all terrestrial plant life as well. The conservation of soil has assumed new dimensions in recent decades as erosion, flooding, and the removal of plant cover have led to widespread losses of fertile soil cover.

Soil scientists classify and survey soil types, study soil microbiology and ecology, and experiment with different methods of soil management, weed control and crop production. Most soil scientists specialise after a general training in one or more of the main life sciences. There are openings in the UK (for example, with the Soil Survey of England and Wales), but the profession is more advanced in the United States. Specialists are also in demand in developing countries.

Maths and Chemistry

Maths is now essential for all management and consultancy work, as is the ability to use computer systems. Chemistry is necessary for most areas of work except for certain types of wardening.

Computer Sciences

Specialised computer science skills are in high demand among environmental institutions and organisations in most sectors. The principal applications are monitoring, database management, and developing early warning or long-term forecasting models.

Qualifications

Anyone starting out on a career in environmental science has two options; to find work straight from school as a technician or laboratory assistant, usually taking qualifying examinations such as BTEC or RSA, or to go on to further education. The minimum requirements for school leavers wanting to start work right away are at least three GCSEs, including one in a relevant science. At least one additional A level would be an advantage.

Courses in life sciences are offered at universities, polytechnics, technical colleges and other centres of higher education throughout the country. General courses are available, and there are many first degrees on offer that specialise. Students must decide whether to take a general mixed discipline degree first, and then perhaps a more specific qualification later, or to specialise first and then take an 'applied' qualification. The normal requirements for entry to a course of higher education are five GCSEs (with at least one in a science subject) and two A levels in science subjects (one of which should relate closely to your chosen course). It is worth noting that sandwich courses offer opportunities for practical work experience and employers usually prefer applicants with such experience.

Openings

There may not be many openings for scientists with individual voluntary, statutory, government, or resource management organisations, but the number employed in all sectors makes conservation an important career option. The problem is that it is also a popular option, and the number of scientists who would like to work in conservation exceeds the number of vacancies.

The largest problem facing environmental scientists is that they are often compelled to compete for jobs that could be filled by candidates with a more general education, as a knowledge of science may be only one of several requirements. A second problem is that courses in ecology and environmental science offered at schools and centres of higher education are relatively new and untried, and are less familiar to employers. Third, many scientists are employed in research, a field susceptible to spending cuts.

On the other hand, there are advantages in having a science background. It can be an advantage, for example, when competing for vacancies in library and information work, writing and publishing, education, museum work, administration and planning.

For the candidate who succeeds in finding work in environmental science, the rewards are many and the openings varied. Research and development, ecology and field work, reserve management, the running of zoos and botanical gardens, and conservation policy are just some of the areas that need specialist knowledge.

Being Prepared

Understanding the profession, knowing where to look for jobs and anticipating demands can save a great deal of wasted time and energy.

Because vacancies are rare you cannot afford to miss many opportunities through lack of preparation.

First, it is important not to specialise too early on in your career. A general education in one or more scientific disciplines will widen your options. Studying other sciences during training can be useful, and you can try to pick up part-time or unpaid practical experience and learn additional skills, such as computer science and data-handling techniques. Use holidays and weekends to go on field trips, undertake your own research projects, work as a voluntary warden, or take classes in related subjects.

Because practical experience is so important, it is advisable to work with local voluntary organisations in school and college holidays, and to become proficient in some aspect of conservation such as management of a particular habitat, identification of one or more groups of organisms, or organising teams of volunteers. Writing for local newsletters helps to develop a good reporting style and all these activities will get you in touch with other people in conservation. Almost all successful conservation scientists started out as amateurs with natural history or conservation as a hobby while they were still at school.

Second, be sure that you know the job market and all likely employers. Cast your net as widely as possible, and explore even remote possibilities. Join any professional societies linked with your science, attend their meetings, become active in their work, and meet as many other members as you can. One of them might point out a vacancy that you have not heard about, or might even be able to offer you part-time or full-time work.

Finally, start looking around and apply for jobs some time before you leave school, college, or university. Many organisations, with their eyes on school leavers or graduates coming on to the job market in the late summer, will start advertising jobs early in the year, in March or April. You should be prepared to compromise and to accept routine work as a lab assistant or technician if it arises. Such work will often give you the kind of useful experience that will help you to move on to something better. You may even be obliged to accept work for a while in a field unrelated to conservation, but with care you can turn such experience to your advantage.

Case Studies

Jane is a biologist with a government research department.

When I decided to read biology at university I didn't really have any idea what I would do with it when I graduated. My careers adviser at school said that jobs would be hard to find, but I didn't realise just how hard until I was about halfway through the course and I had to start thinking about jobs. I was in touch with past graduates who seemed to be either unemployed or doing completely unrelated work. I was keen on the idea of working in conservation, but I soon found out that I had chosen the most competitive area of all, and that most of the decisions in conservation are taken by managers, politicians and planners, rather than scientists.

My first job was with a science publishing firm, where I was really no more than a glorified secretary. I stuck that for about seven months, and then I was lucky enough to get a job with a pharmaceutical company as a lab assistant, which actually gave me useful experience. I spent a year there, meanwhile applying for every job I could find in the environmental field. I finally found this job, and I only got it because it is a very junior post that would normally be filled by a school leaver with four GCSEs. In a way I feel exploited because I'm being paid a pittance, much less than my qualifications deserve, but it seems the only way to do the work I want to do. I'm involved in research, and, while the work can be a bit routine at the moment, at least I have my foot in the door. Fortunately, I work with a good team of people who are giving me more responsibility than someone in this post would normally have, so for the first time since graduating I'm quite confident.

Paula is a scientific officer with a statutory conservation body, a job that she found soon after graduating three years ago.

I am the first to admit how lucky I am, and I could hardly believe it when I got this job because it was only the third that I had applied for, and it's exactly what I want. My post involves representing the organisation out in the field, which means a combination of public relations and politics. I travel around meeting private landowners, farmers, foresters, planners, and local government officials to talk about land management and to iron out any problems or conflicts of interest over our various protected areas. It can often be frustrating and bureaucratic, especially when you're dealing with public inquiries into threats to protected land, but, at the same time, it is very satisfying to know that you are putting conservation policy into practice. It is by no means a desk job, and I'm able to use my degree to good effect.

Chapter 9

Creative and Media Opportunities

Introduction

Now that the principles of environmental conservation are firmly established in law, the challenge remains to enshrine them in custom, to win the hearts and minds of ordinary people everywhere round to the wisdom of caring for the environment and enlist their wholehearted support for and participation in sustainable development. Here is a task that requires the special abilities of creative and media professionals to 'reinvent' nature and inspire and motivate others to take a pride in their environment.

Film-making

Britain has a long tradition of craftsmanship in wildlife film-making. A measure of the ability of many wildlife film-makers is that they make their work look so effortless and accessible that film-making seems easy. Nothing could be further from the truth.

Not only is the field competitive, but the work is demanding and arduous. The pleasure of working with film and of seeing a project through to completion is tempered by long hours of hard and sometimes tedious labour. Film-making is a career in which there are no easy options and no quick ways to the top.

That said, though, there are many people making wildlife films, and new film-makers will be needed in future. Jobs are available for those with very good qualifications.

Learning the Craft

Film making involves both artistic and technical skills, and you should be sure that you possess these skills. The best way of getting started is to watch and absorb as many films as possible, noting the quality of the visual effects and how they were achieved - the editing, camera movements, lighting, sound effects, use of an on-screen or off-screen narrator, and so on. Fortunately, wildlife films are common enough on television to make such a study possible. You will find the best films spring from passion and commitment, not just the recycling of facts. Visual flair and a good storytelling sense are more important than science qualifications every time.

At the same time, you should be experimenting with making films of your own. For this you will need either a good 9mm camera or a 16mm camera. Working with 16mm may be more expensive, but it brings you closer to the kind of situations and problems faced by professional film-makers, who work largely in that format. If possible, attend film-making classes and learn the theory from some of the many practical film-making guides that are available.

Once you understand the principles of film-making, you have the choice of either formal training at a college or film school (where you learn film-making as a craft and would be left to specialise in wildlife film-making in your own time) or a full-time job. Formal training is no guarantee of a job, but it does help because you can assess the job market from the inside. For details of long- or short-term courses in film-making, see *Media Courses UK*. Also read *Careers in Film and Video* (Kogan Page).

Finding a Job

To find a job you will have to prove your ability, and few employers will consider you unless you have some of your own work to show. For that reason it is worth undertaking as many exercises and projects as you can, either alone or through your training centre. Keep filming and experimenting until you are sure that you have something worth showing. Be very critical of your work and be prepared to shoot a lot of wasted footage before you have a scene, a sequence or a complete exercise worth showing. Only a prodigy can come up with a winner first time around.

The job market is largely limited to television companies and independent film production companies. For television work the choice at the moment is BBC, ITV or Channel 4, and the normal point of entry is a post as a trainee in whatever field interests you. It could be editing, sound, lighting, camerawork, or working as a production assistant. Both BBC and ITV regularly recruit new trainees, but for every vacancy there are likely to be at least 100 applicants. Those with proven interests and proven skills, but not necessarily experience, stand the best chance.

Once accepted by a television company it is a matter of working your way up through the normal internal promotion structure, proving your ability and expressing your interests. In most instances you should anticipate a long apprenticeship.

If you opt for work with a film production company, you have many more smaller organisations from which to choose. It may mean starting out in a menial or very junior post, and finding where the openings are likely to occur is a matter of keeping your ear to the ground. There are very few film production companies that specialise in wildlife film-making, but the field is changing all the time which means that opportunities vary.

Because the field is both changeable and competitive, it is difficult to generalise about careers in film-making. Although several thousand people are employed in film-making in the UK, few specialise in wildlife and these will take on other work as well. To succeed in film-making you should prove your abilities, be persistent and resilient, and enjoy a great deal of good luck.

Photography

Like film-making, photography demands both technical and artistic aptitudes, and although it is competitive it offers many more openings and outlets than film-making. Film-makers are restricted to working for television or film production companies, while photographers can sell their work to a host of magazines and journals, and to photo libraries, audo-visual companies, and galleries. It is also much easier and cheaper to set up with the right equipment. However, photography demands both ability and initiative.

Learning the Craft

Studying the work of other photographers, learning their techniques and carrying out your own experiments are essential. The basic tools are a good reflex camera (either 35mm or 2¼″ square) and a range of lenses: at least a standard lens (50mm) and preferably a wide-angle lens (28mm), a medium telephoto (90-135mm) and a telephoto lens (200-600mm). As you become more specialised you will need specialist accessories such as remote controls, flash units, automatic winders, micro-focus lenses and extreme telephoto lenses.

You can teach yourself the basics of photography with advice from any of the many practical manuals and guides that exist. Otherwise, there are a number of part-time and full-time courses available at polytechnics, colleges of art and universities. All the courses will teach you the craft, but qualifications do not guarantee you employment.

Finding a Job

Photographers have two basic options: to work in either full-time or freelance employment. Full-time work is very difficult to find for the general photographer but even more so for someone wanting to specialise. A handful of journals employ wildlife photographers, but most buy photos in from freelances or photo libraries. A more realistic option is to work in a general capacity and pursue wildlife photography as a side-line or as just one aspect of the job.

Freelance work offers many more opportunities, but demands initiative, resilience and the ability to work alone. Most freelances sell their work through photo libraries or try to get commissioned work with a journal, a publisher, or an organisation that needs photographic material. It generally takes a great deal of time to become established, and for every photograph you sell you will have many rejected. You might lodge a large collection of photographs with a photo library and never make a sale, although you can guard against this by asking them which subjects are in most demand and where they have gaps that need filling.

It is difficult to make a living out of photography, particularly wildlife photography, but many manage to succeed, so the opportunities are there.

Writing

Writing is a creative skill that has a longer and more developed heritage than either film-making or photography. To write successfully means you must have something to say and be able to express it in a succinct and effective manner. Conservation relies heavily on the printed word to convey the issues involved, and offers outlets for writers in journalism, information, education, press and public relations, and publicity.

Learning the Craft

Much can be learned from studying and imitating other writers. Writing well means being able to inform, entertain, educate, and even inspire readers, and demands not only an understanding of your topic but a wide general knowledge. To write well, you must have an interest in everything happening around you. Learning the skill means reading and writing copiously. For the environmental writer, it means understanding all the issues and knowing how to convey them to those who know about the subject and to those who do not.

An advantage is that the basic tools of the trade are cheap. Apart from paper and a word processor, many writers manage to exist with only a good dictionary, a thesaurus, and a copy of the *Writers' and Artists' Yearbook.*

Finding a Job

There are outlets for environmental writers in both the printed media (books, newspapers and journals) and the broadcast media (radio and television). Almost any organisation that wants to convey information will need to employ writers, either on a full-time or a freelance basis.

Full-time writers find a variety of work: writing for a periodical, running a press office, writing and publishing information material, putting together technical and practical manuals, or managing a press relations programme. Almost all the organisations mentioned in this book use writers in one capacity or another.

The freelance environmental field is already well subscribed, but there is always room for new writers. Making a living out of freelance work is not easy, especially for someone just starting out. Becoming established demands patience, resilience and application, and staying established demands a knowledge of the marketplace and the ability to offer a service. Most freelances derive their main income from regular commissioned work which comes from one or two main sources, and make up the balance from whatever else they can sell.

Case Studies

Sandy has spent the last three years working as a librarian in a photographic agency that specialises in wildlife subjects. It is less of a career for her than a regular job that allows her to pursue freelance wildlife photography in her own time. She has been taking photographs since she was five, encouraged by her father who owned his own photographic studio. She was regarded as unusual at school because there weren't many girls interested in photography as a hobby, let alone as a career. Her first job after leaving school was in her father's studio, after which she worked as a sales assistant in a camera shop, a job that didn't appeal to her.

> I had been trying to sell some of my work through various photo agencies, and made a couple of friends at the agency where I now work. They told me there was a vacancy for a librarian. I applied, and here I am. It's actually more fun than it sounds. I spend most of my time cataloguing photographs and sending selections out to clients, but I also work with wildlife photographers, going through their work and commissioning photographs. I get to talk to a lot of them about different techniques, and I try to get out every weekend and most evenings during the summer to take photos. I go for everything from general landscapes and habitats to plants, trees, flowers, mammals and birds. I have been experimenting recently with remote control flashlit photography of animals like badgers and pine martens, which has been an experience. I am gradually selling more and more of my work: to books, magazines, newspapers, and advertising and PR agencies. My greatest ambition is to have a book of my work published, but that's likely to be a few years off yet!

Chris is the assistant press officer for a national conservation charity. He began his working career after dropping out of art college and being offered a job as a copywriter with an advertising agency. It was a stop-gap job to give him time to think about a career. While he was copywriting he wrote to the charity for which he now works and offered to help them on a voluntary basis.

> I'd always been interested in their work and I just wanted to help somehow. They commissioned a few stories from me for their members' magazine, and when this job came up I applied and was lucky enough to get it. My copywriting experience helped a lot. I am enjoying the job enormously, especially as it gives me time to do a lot of freelance writing in the evenings and at weekends. I'm virtually running the press programme, writing press releases, features, leaflets, information packs, and the occasional audio-visual programme, as well as setting up press conferences and trying to get our activities covered in the press. My plan is to go freelance eventually, and live off writing and maybe a bit of photography. It's still too early, though, because you have to build up the contacts first. At the moment I'm concentrating on writing as much as I can and building up a portfolio.

Green Art and Design

An unusual growth area in creative opportunities relevant to environmental concerns, green art and design can involve anything from the creation and management of art trails and sculpture parks in protected areas, 'clean' product and packaging design for industry, community-based public art initiatives in inner cities and rural areas, interpretive drawing for science-related purposes such as botanical or natural history illustration and much besides.

A green design course is available at undergraduate level in West Surrey College of Art and Design and a few other art schools while the School of Art at the University of Brighton runs elective courses in environmentally relevant public art. A postgraduate research degree course in Art and Environment is on offer at Manchester Metropolitan University. A natural history illustration course at the Royal College of Art leads to an MA. Botanical drawing courses are also being developed in collaboration with the Royal Botanical Gardens at Kew.

These are but a few indicative examples of educational trends reflecting new environment-related activities currently astir in creative domains outside the mass media. Active art groups such as Common Ground and Projects Environment have been engaged in art and environment projects at community level for some time now and, though not as much a part of the art scene as they are, for example, in many parts of the USA, these activities are rapidly coming of age in Britain and Europe.

General Guidance

Chapter 10
Working Overseas

Introduction

Because nature is not divided by national boundaries, there is a limit to how much individual nations acting alone can achieve in conservation. Birds migrate across the world, forests spread across many countries, rivers flow across continents, and oceans cover the planet. The creation of so many international conservation organisations over the past 30 years is some indication of how important international co-operation has become.

The need is particularly evident in the developing countries of the global South, where governments face the greatest problems in balancing the conservation of their natural resources against the economic and social needs of their people. Most have tried to condense industrial revolutions that took hundreds of years in Europe into only a few decades, often with unforeseen and unfortunate side-effects.

Most Southern governments are now heavily in debt to the industrialised North for development aid in the form of capital, materials and education, and that aid has not always been appropriate to local needs. Dams have been built without the necessary research into river flow characteristics; forests have been cleared to make way for agricultural projects unsuited to local soil and climatic conditions; improved medical care without parallel family planning has contributed to massive population growth; unequal land distribution has put too much pressure on agricultural land. The natural environment has suffered as a consequence, and the threat posed to tropical animal and plant species is just one of the negative results.

All this has meant that many of the most pressing environmental issues are to be found in the South, and temporary contract work or even permanent settlement abroad has many attractions for conservationists and environmental planners from the North.

Working with a British Organisation

There are a number of British or British-based organisations that either study the problems of other countries or run active environmental and development aid projects abroad. Charities such as Oxfam, WWF (World Wide Fund For Nature, see Chapter 3) and the Fauna and Flora Preservation Society raise money for conservation projects abroad, and bodies such as the Overseas Development Administration, particularly its scientific unit, the Natural Resources Institute (see Chapter 5) concern themselves with conservation or agronomy aspects of Third World development issues. Many aid and natural resource management programmes are run through the Commonwealth and similar co-operative bodies. Environment and development policy research is undertaken by the International Institute for Environment and Development and the Overseas Development Institute.

Before approaching any of these bodies it is important to distinguish between those that offer openings within the United Kingdom only, and those that employ staff for overseas postings. Jobs that involve overseas travel are rare, very much in demand, and usually go only to candidates with a high level of relevant experience and qualifications.

Working with an International Organisation

There are a number of international organisations, based outside the UK, that conduct most of their business in English and frequently recruit people from the UK. These include United Nations bodies, such as the UN Environment Programme, the Food and Agriculture Organisation and the World Bank, non-governmental organisations such as the World Conservation Union (IUCN) and the World Wide Fund For Nature, based in Switzerland. Other international agencies of interest to environmentalists are listed in Chapter 14.

Vacancies in international organisations are not only highly sought-after, but are competed for by applicants from all over the world and the standards expected of them are generally high. You will need substantial relevant experience and qualifications, and it helps if you can speak another language (usually French or Spanish) and have lived or worked abroad before.

Working with a Voluntary Aid Agency

Taking part in voluntary aid programmes is an effective method of helping Third World communities learn to meet their economic and social needs, and therefore to manage and protect their environment. Britain currently sends about 800 volunteers overseas every year to some 50 countries. They are sent under the auspices of a variety of voluntary agencies: the four best known are the Catholic Institute for International Relations, Voluntary Service Overseas, Skillshare Africa (formerly International Voluntary Service) and the United Nations Association International Service.

Volunteers generally need to have useful skills and qualifications, mostly in agriculture, education, health services and crafts. Almost half the volunteers sent abroad are teachers. Skills needed by VSO in the field of natural resources are those of: agriculturalists, agronomists, farm managers, community agro/foresters, agricultural scientists, environmentalists, marine and freshwater fishery experts, horticulturalists and livestock specialists, as well as town planners and surveyors.

Applicants should be suitably qualified, aged over 21, in good health, and be willing to work for a minimum of two years at subsistence rates of pay or on a salary based on local rates. In the case of the Catholic Institute for International Relations, a UK allowance in addition to the overseas salary is also available, as well as free accommodation, flights, insurance, language training and a pre-departure grant. In general, only married couples with no dependent children are accepted, but VSO is now taking volunteers with dependants in certain skill shortfall areas. Details in *VSO and Dependants* (for address, see p. 112). Recruitment is arranged nationally, and most of the agencies will supply applicants with information about current vacancies.

Many projects are aimed at increasing food production, improving community health, applying family planning, and helping rural communities to become self-sufficient. This kind of work provides anyone interested in conservation with useful experience and a perspective on problems faced by the Third World.

Undertaking Research Projects

A number of government and voluntary bodies run research programmes overseas, generally in a branch of life science or on some aspect of development aid. Most postings go to graduates in a related field who are already employed with the sponsoring organisation. The postings vary in their nature and duration; a project may last only a few weeks or, in the case of some British Antarctic Survey postings, can last three years, and demand the resilience to cope with living in close proximity to a small group of researchers for considerable periods in isolated and physically demanding conditions.

Alternatively, if you are lucky and have the right skills and qualifications, you could be invited to join a one-off conservation project or research expedition.

Emigrating or Settling Abroad

If you plan to travel or settle abroad for any length of time, the nature and scope of the options available to you change. Young people without particular skills, but with initiative, have been known to travel abroad and find work as rangers or guides in national parks, as members of local conservation groups, or as short-contract workers.

Planners and scientists with professional qualifications may find it relatively easy to be accepted as immigrants by countries such as Australia, Canada and New Zealand, where there are many openings in the environmental field, notably in national park and reserve management. Similarly, Third World countries often take on resource planners and managers for consultancy or contract work.

Perhaps the greatest number of career and further training opportunities are to be found in the United States. Resource management, particularly where it applies to forestry, soil, watershed, wildlife and range management, has a constant need for trained personnel, although the field is highly competitive.

Chapter 11

Finding and Applying for Jobs

Introduction

Environmental conservation has become such a popular career choice that there are not enough jobs to go around, and the competition for vacancies is intense. Anyone who has tried writing to one of the many voluntary bodies, and has received a reply, will be only too familiar with such warnings.

On the other hand, it must be said that there *are* vacancies, that *someone* has to fill them, and that the well-prepared candidate is more likely to succeed than any other. One of the biggest problems that newcomers have had to face until now is the lack of accurate and informed career guidance. Conservationists themselves are often unaware of the range of careers the profession offers, and generally find it hard to offer any advice.

Looking for a job in conservation demands, above all, initiative. Remember that you are entering a profession where precedents are still being set, which can make it difficult to learn much from the examples of others who have gone before. It is still at the stage where the newcomer often has to make his or her own opportunities.

You will also need persistence, determination, and luck. Luck you cannot anticipate, but a few fundamental precautions will help you to find the job you want.

Know the Field

Many opportunities are missed because people do not fully understand what conservation involves and do not know what kind of people conservationists are. Newcomers tend to head instinctively for the wildlife organisations, because they are the best known and attract the most publicity. It is often a good idea to cast your net much wider and to look at jobs that involve other kinds of resource management and conservation. They all link up with one another in the end.

Know Your Interests

Once you have learned more about conservation you will be in a much better position to know which aspects interest you. Try not to specialise too early. Be prepared to consider any offers that come your way, because even the dullest post can teach you more about the job and can lead to better things.

Make Yourself Valuable

In our competitive world, specialist knowledge counts for a great deal, and even the most humble qualification can tip the scales in your favour if it makes you better than your competitors. You might know how to use a chainsaw or how to programme a computer. You might have a GCSE in zoology, a certificate earned from taking evening classes at a local college, or a university degree. It all helps.

Above all, though, you need experience, and conservation has the advantage over many other popular careers of offering unlimited opportunities for picking up voluntary unpaid experience. For just a few hours of your time you could learn how to manage a nature reserve, carry out a bird or mammal survey, run a publicity campaign, raise money, and more. Such experience will make you more familiar with conservation and more competitive when it comes to applying for a full-time job.

Know Where to Find the Jobs

It does no harm to send a letter to a conservation organisation asking about possible vacancies and enclosing your cv (and an sae), but most organisations receive many such letters. They rarely pay off, and many go unanswered altogether.

It makes much more sense to join some of the organisations that interest you and become actively involved, to subscribe to selected magazines and periodicals that are not available in your local library, and to keep an eye on all the newspapers and journals that carry job advertisements. The list of journals in Chapter 13 gives a general guide to sources of information on different careers and related vacancies.

Unfortunately, not all vacancies are advertised in the public domain, and some appointments are made internally. Joining conservation groups and meeting conservationists is the best way of keeping your ear to the ground.

Be Prepared and Persistent

You may land the ideal job on your first try, or you may go through endless applications without success, but with preparation and persistence the job you want must sooner or later come your way. Work carefully on your cv, putting in everything you think is relevant without being trivial, and presenting your case thoroughly, logically and neatly. Accept that you will have rejections, but turn each one to your advantage by learning from it and going into each successive application better prepared.

If you approach the job market realistically, stay aware of the competition, and try to strike the right note, you are bound to succeed eventually.

Future Prospects

New conservation bodies are being set up so frequently that directories of environmental groups are out of date almost before they are published. Governments are setting up new departments and research units to manage natural resources and monitor environmental laws. Training in conservation is now more widely available, and research and discussion are teaching us more about the natural world and the interdependence of life on earth.

The demand for jobs may be outstripping the opportunities at the moment but, as environmental management becomes more widely applied, so the openings will grow. The overall forecast is for growth in the scope and application of environmental conservation, greater professionalism in approaching the issues, and a corresponding growth in the number of career opportunities.

Reference

Chapter 12

University and College Courses and Qualifications

The best place to start if you want to find out what environmentally relevant courses are available in tertiary schooling and beyond, is to check the ECCTIS database, accessible through careers offices and some Jobcentres or larger libraries. ECCTIS also includes distance-learning courses and many institutions offer shorter courses for continuing professional development. You will find environmental science courses widely available but others, say countryside management or environmental law, less common.

Many universities are beginning to build environmental factors into every part of the curriculum, in response to the recommendations of the 1993 Toyne Report, which urged an all-round 'greening' process as distinct from a compartmentalised view of environmental knowledge as a subject apart. It should soon be possible for students to find the best course they can in any career field they choose, then press hard for maximum environmental content within it. WWF's Education Division provides three-day courses for university staff interested in developing environmental modules across the disciplines.

Dartington Hall is building a reputation for residential courses which develop the philosophical base to environmental action. Many colleges have environmental courses in their night school, part-time degree facilities or Continuing Education programmes.

The Environment Council publishes a biennial *Directory of Environmental Courses*, sponsored by the Department of the Environment, which also lists many useful further sources of information. The range of subject areas the directory includes is listed on page 90.

The Open University offers distance-learning courses, among which the most useful are 'The Changing Countryside' and the 'Practical Conservation' series. The Field Studies Council provides courses in Countryside Management and in Biological Surveying: these are parts of certified courses run in conjunction with Otley College of Agriculture and Horticulture.

London University Extramural Department certificate and diploma courses in Field Biology, and in Ecology and Conservation are correspondence courses combining field-trips, home-based projects and a final examination.

For information about vocational training opportunities in environmental and resource management, see the *Work Experience* section at the end of Chapter 3 of this handbook.

Topic Areas Listed in the Environment Council Directory

Agriculture and fisheries
Archaeology
Architecture
Building and housing
Civil engineering
Development
Ecology
Environmental economics
Environmental education
Environmental law
Environmental management
Environmental sciences
Forestry
Horticulture and garden design
Industry and environment
Land management and conservation
Landscape architecture and design
Town and country planning
Urban studies

This chapter suggests a few course and subject ideas - some specialised, some more general. It doesn't list courses in journalism, law, geography, agriculture, architecture, building, engineering, ecology, science or development studies, but these - and others - may often be just as useful. Before deciding which course suits you best, it is advisable to find out more about what each entails from the university or college concerned. Further details of courses are given in the *UCAS Handbook*, the *Directory of Environmental Courses* from The Environment Council, 21 Elizabeth Street, London SW1W 9RP; *Environmental Careers Handbook*, published by Trotman; and *Careers and Courses in Sustainable Technologies*, published by CAT Publications.

Countryside Conservation and Management

University of Aberdeen, Department of Agriculture, 581 King Street, Aberdeen AB9 1UD
3- or 4-year BLE/BLE (Hons) in Land Economy
3-year BSc in Marine Resource Management
4-year BSc (Hons) in Countryside with Environmental Management
1- or 2-year MSc/PgDip in Rural and Regional Resources Planning

Aberystwyth: The University College of Wales, Aberystwyth, Dyfed DY23 2AX
3-year BSc (Hons) in Rural Resource Management

Anglia Polytechnic University, Writtle College, Chelmsford, Essex CM1 3RR
3-year BSc (Hons) Rural Resource Development
HND in Rural Resource Management with option in Habitat Conservation

Askham Bryan College of Agriculture and Horticulture, Askham Bryan, York YO2 3PR
1-year BSc Land Management and Technology

Bangor: University of Wales, Bangor, Gwynedd LL57 2DG
3-year BSc (Hons) in Rural Resource Management
1-year MSc/Diploma in Rural Resource Management

Bishop Burton College, Bishop Burton, Beverley, North Humberside HU17 8QG
HNC in Rural Studies (Countryside Management)
3-year BSc (Hons) in Countryside Management

Bournemouth University, Poole House, Talbot Campus, Fern Barrow, Poole, Dorset BH12 5BB
3-year BSc/BSc (Hons) in Environmental Protection

University of Buckingham, Buckingham MK18 1EG
3-year BSc (Hons) in Agriculture and Land Management

University of Cambridge, Cambridge CB2 1TN
3-year BA (Hons) in Land Economy
2- or 3-year MLitt/MSc/PhD in Land Economy

Cardiff: University of Wales, College of Cardiff, PO Box 68, Cardiff CF1 3XA
1-year MSc/Diploma in Technical Change and Regional Development

Coventry University, Priory Street, Coventry CV1 5FB
4-year BSc (Hons) in Recreation and the Countryside
1–2½ year MA/Diploma in Regional Planning

Cranfield University, Silsoe College, Silsoe, Bedford MK45 4DT
1/2 year MSc in Drainage and Land Reclamation Engineering
1/2 year MSc in Engineering for Rural Development
1-year MSc in Land Resource Management
1-year MSc in Range Management
1-year MSc in Soil Conservation
9-month PgDip in Land Resource Planning

Cranfield University, Shuttleworth College, Old Warden Park, Biggleswade, Bedfordshire SG18 9DX
2-year HND in Rural Environment Management

Doncaster College, Waterdale, Doncaster DN1 3EX
3-year BSc in Environment and Resource Management

University of Dundee, Perth Road, Dundee DD1 4HN
LLM/PhD in Natural Resources Law and Policy
1 year LLM/Diploma in Natural Resources Law and Policy

University of East Anglia, Norwich, Norfolk NR4 7TJ
1-year MSc in Land Use Planning
1-year MA in Rural Development

University of East London, Barking Campus, Longbridge Road, Dagenham, Essex RM8 2AS
3-year BSc/BSc (Hons) in Wildlife Conservation
5-year BSc in Land Administration

Edge Hill College of Higher Education, Ormskirk, Lancs L39 4QP
3-year BSc (Hons) in Field Biology and Habitat Management

University of Edinburgh, Edinburgh EH8 9YL
4-yeare BSc (Hons) in Agriculture, Forestry and Rural Economy
1-year MSc/Diploma in Resource Management, Forestry, Agriculture and Ecology

Farnborough College of Technology, Boundary Road, Farnborough, Hampshire GU14 6SB
3-year BSc (Hons) in Environmental Protection

University of Glasgow, Glasgow G12 8QQ
1-year MSc/Diploma in Water Resources Engineering Management

Harper Adams Agricultural College, Newport, Shropshire TF10 8NB
4-year BSc (Hons) in Rural Environmental Protection
4-year BSc (Hons) in Rural Enterprise and Land Management

Herefordshire College of Agriculture, Holme Lacy, Hereford HR2 6LL
2-year HND in Countryside Recreational Management
2-year HND in Leisure Studies: Countryside Management

Heriot-Watt University, Riccarton, Edinburgh EH14 4AS
1-year MSc/Diploma in Marine Resource Management
1-year MSc/Diploma in Marine Resource Development and Protection

University of Hertfordshire, College Lane, Hatfield, Hertfordshire AL10 9AB
2-year HND in Landbased Industry

University of Hull, Hull HU6 7RX
3-year BSc (Hons) in Environmental and Resource Management

King's College, Strand, London WC2R 2LS
1-year MSc in Aquatic Resource Management

Kingston University, Penrhyn Road, Kingston upon Thames, Surrey KT1 2EE
3-year BSc/BSc (Hons) in Resources and the Environment

Lackham College, Lacock, Chippenham, Wiltshire SN15 2NY
2-year HND in Rural Studies (Game and Wildlife Management)

University of Leicester, University Road, Leicester LE1 7RH
1-year MSc/Diploma in Natural Resource Management

Liverpool John Moores University, Great Orford Street, Liverpool L3 5YD
3-year BSc/Hons in Countryside Management

University of Luton, Park Square, Luton, Bedfordshire LU1 3JU

3-year HND in Land Administration (Geographical Techniques)
2-year HND in Land Administration (Estate Management)

University of Manchester, Oxford Road, Manchester M13 9PL
3-year plus BA (Hons) in Landscape Planning and Management
1-year MA (Econ) in Economics and Management of Rural Development
1-year Advanced Diploma in Rural Development Education

Napier University, 219 Colinton Road, Edinburgh EH14 1DJ
3-year BSc/BSc (Hons) Rural Resources
HND Rural Resources

University of Newcastle upon Tyne, Newcastle upon Tyne, NE1 7RU
3-year BSc (Hons) in Countryside Management
3-year BSc (Hons) in Natural Resources

Normal College of Higher Education, Bangor, Gwynedd LL57 2PX
1-year PgDip in Countryside Management

North East Wales Institute of Higher Education, Newi Plascoch, Mold Road, Wrexham, Clwyd LL11 2AW

University of Nottingham, University Park, Nottingham NG7 2RD
3-year BEng (Hons) in Environmental Engineering and Resource Management
1-year MA/Diploma in Heritage Studies

University of Oxford, University Offices, Wellington Square, Oxford OX1 2JD
1-year MSc in Forestry and its Relation to Land Use

University of Plymouth, Seale-Hayne Campus, Newton Abbot, Devon TQ12 6NQ
HND in Countryside Recreation Management
4-year BSc (Hons) in Agriculture and Countryside Management
4-year BSc (Hons) in Rural Resource Management
4-year BSc (Hons) in Rural Estate Management

Faculty of Urban and Regional Studies, University of Reading, Whiteknights, Reading RG6 2AH
3-year BSc (Hons) in Rural Land Management
4-year BSc (Hons) in Landscape Management
3-year BSc (Hons) in Rural Resource Management
3-year BSc (Hons) in Land Management
1-year MSc in Land Management
3-year MBA in Food and Rural Management
1-year Diploma in Rural Extension and Women
1-year MA/Diploma in Rural Social Development
1-year MSc/Diploma in Wildlife Management and Control

Royal Agricultural College, Cirencester, Glos GL7 6JS
3-year BSc (Hons) in Agriculture and Land Management
3-year BSc (Hons) in Rural Land Management
3-year Professional Diploma in Rural Estate Management
1-year MSc/PgDip in Rural Estate Management
1-year MA in Rural Policy Studies

The Scottish Agricultural College (Edinburgh, Aberdeen, Ayr), Academic Registry, Freepost, Ayr KA6 5HW
3/4-year BSc/BSc (Hons) in Rural Resources
HNC/HND in Countryside Recreation and Conservation Management
HND in Rural Resources

University of St Andrew's, College Gate, St Andrew's, Fife FY16 9AJ
1-year MPhil/MLitt/MSc/Diploma in Land Resources and Land Utilisation

Sheffield Hallam University, Pond Street, Sheffield S1 1WB
1-year MSc/PgDip/PgCert in Countryside Recreation Management

University of Southampton, Highfield, Southampton SO9 5NH
1-year MSc/Diploma in Soil Conservation and Land Reclamation Engineering

Stoke-on-Trent College, Couldon Campus, Stoke Road, Shelton, Stoke-on-Trent ST4 2DG

University of Sussex, Sussex House, Falmer, Brighton BN1 9RH
1-year MA in Rural Development

University of Wales, Singleton Park, Swansea SA2 8PP
1-year MSc in Regional Development Planning

University College London, Gower Street, London WC1E 6BT
1-year MSc/Diploma in Conservation

University of the West of England, Frenchay Campus, Coldharbour Lane, Bristol BS16 1QY
3-year BSc (Hons) in Environmental Quality and Resource Management

University of Wolverhampton, Molineux Street, Wolverhampton WV1 1SB
1-year MSc in Rural Development and Training

Wye College, University of London, Ashford, Kent TN25 5AH
3-year BSc (Hons) in Countryside Management

Yeovil College, Ilchester Road, Yeovil, Somerset BA21 3BA
1-year Foundation Environmental Protection

The University of York, Heslington, York YO1 5DD
3-year BSc (Hons) in Ecology, Conservation and Environment
3-year BSc (Hons) in Chemistry, Resources and the Environment

Environmental Management

University of Aberdeen, Department of Agriculture, 581 King Street, Aberdeen AB9 1UD
1-year MSc/Diploma in Public Health

Aberystwyth: The University College of Wales, Aberystwyth, Dyfed DY23 2AX
2-year MSc/Diploma in Environmental Auditing (distance learning)
1-year MSc/Diploma in Environmental Rehabilitation
1-year MSc/Diploma in Environmental Impact Assessment

Anglia Polytechnic University, Writtle College, Chelmsford, Essex CM1 3RR
1-year MSc in Environmental Assessment

Bath College of Higher Education, Newton Park, Newton St Loe, Bath BA2 9BN
2-year DipHE in General Studies
3-year BA/BSc (Hons) in Combined Studies

University of Bath, Claverton Down, Bath BA2 7AY
4-year BEng/MEng in Chemical Engineering and Environmental Management
1-year MSc/Diploma in Environmental Science, Policy and Planning

University of Birmingham, Edgbaston, Birmingham B15 2TT

2-year MSc in Environmental Health
1-year MSc in Public and Environmental Health Sciences
1-year MMed Sc/PhD in Public Health and Epidemiology

Bournemouth University, Poole House, Talbot Campus, Fern Barrow, Poole, Dorset BH12 5BB
3-year BSc (Hons) in Environmental Risk Management
1-year MSc/PgDip in Coastal Zone Management

University of Bradford, Richmond, Bradford BD7 1DP
3-year BSc (Hons) in Environmental Management and Technology
1-year MSc/PgDip in Business Strategy and Environmental Management
1-year MSc/PgDip in Environmental Monitoring

University of Brighton, Lewes Road, Brighton BN2 4AT
1-year MSc in Environmental Impact Assessment

University of Bristol, Senate House, Tyndall Avenue, Bristol BS8 1TH
2-year MSc Dip/Cert in Ecology and Management of the Natural Environment

Brunel University, Uxbridge, Middlesex UB8 3PH
4-year ENgD in Environmental Technology
1-year MSc in Environmental Science with Legislation and Management

Buckinghamshire College, Queen Alexandra Road, High Wycombe, Buckinghamshire HP11 2JZ
3-year BA (Hons) in Environmental Design
3-year BA (Hons) in Business Administration with Environmental Management

University of Cambridge, Cambridge CB2 1TN
1-year Diploma in Public Health

Cardiff Institute of Higher Education
4-year BSc (Hons) in Environmental Health

University of Wales, College of Cardiff, PO Box 68, Cardiff CF1 3XA
3-year BSc (Hons) in Ecology and Environmental Management

University of Central England in Birmingham, Perry Barr, Birmingham B42 2SU
3-year BSc (Hons) in Environmental Planning
1-year PgDip in Environmental Protection, Control and Monitoring
2-year MA/PgDip in Environmental Management

University of Central Lancashire, Preston, Lancashire PR1 2HE
3-year BSc (Hons) in Environmental Management

Cheshire College of Agriculture, Reaseheath, Nantwich CW5 6DF
3-year BSc (Hons) in Food and Environmental Management

Colchester Institute, Sheepen Road, Colchester, Essex CO3 3LL
3-year HND in Environmental Monitoring and Protection
4-year BSc (Hons) in Environmental Monitoring and Protection

Cranfield University, Silsoe College, Silsoe, Bedford MK45 4DT
1-year MSc in Environmental Water Management

University of Derby, Kedlestone Road, Derby DE3 1GB
3-year BSc (Hons) in Environmental Monitoring

Doncaster College, Waterdale, Doncaster DN1 3EX
3-year BSc in Environmental and Resource Management

Duncan of Jordanstone College of Art, 13 Perth Road, Dundee DD1 4HT
4-year BDes (Hons) in Interior and Environmental Design
4-year BSc (Hons) in Environmental Management

University of Dundee, Perth Road, Dundee DD1 4HN
4-year BSc (Hons) in Environmental Monitoring
4-term MSc in Environmental Health
4-term MPH in Public Health

Durham College of Agriculture and Horticulture, Hunghall, Durham DH1 3SG
2-year HND in Environmental Management

University of East Anglia, Norwich, Norfolk NR4 7TJ
1-year MSc in Agriculture, Environment and Development

University of Edinburgh
1-year MSc/Diploma in Environmental Health

Farnborough College of Technology, Boundary Road, Farnborough, Hampshire GU14 6SB
1-year MSc/PgDip/PgCert in Environmental Management

University of Glamorgan, Llantwit Road, Treforest, Pontypridd, Mid-Glamorgan CF37 1DL
3-year BSc (Hons) in Manufacturing and the Environment

University of Glasgow, Glasgow G12 8QQ
1-year MPH in Public Health

University of Greenwich, Wellington Street, London SE18 6PF
3-year BSc/BSc (Hons) in Environmental Control

University of Huddersfield, Queensgate, Huddersfield HD1 3DH
3-year BSc/BSc (Hons) in Environmental Analysis

University of Hull, Hull HU6 7RX
1-year MSc/Diploma in Environmental Policy and Management

University of Humberside, Milner Hall, Cottingham Road, Hull HU6 7RT
3-year BSc (Hons) in Food and Environmental Management

Imperial College of Science, Technology and Medicine, London SW7 2AZ
1-year MSc/DIC in Environmental Analysis and Assessment
1-year MSc/DIC in Environmental Analysis

Lancaster University, Lancaster LA1 4YW
3-year BSc (Hons) in Environmental Management
1-year MSc in European Environmental Policy and Regulation

University of Leeds, Leeds LS2 9JT
3-year BSc/BSc (Hons) in Environmental Management
3-year BA (Hons) in Environmental Management

University of Leicester, University Road, Leicester LE1 7RH
1-year MSc/Diploma in Natural Resource Management

Liverpool John Moores University, Rodney House, 70 Mount Pleasant, Liverpool L3 5UX
3-year BA/BA (Hons) in Corporate Environmental Management

University of Liverpool, PO Box 147, Liverpool L69 3BX
1-year MPH in Public Health

London Guildhall University, 117–119 Houndsditch, London EC3A 7BU
3-year BA/BA (Hons) in Environmental Policy and Management

London School of Economics and Political Science, Houghton Street, London WC2A 2AE
1-year PgDip in Environmental Management

Loughborough University of Technology, Loughborough, Leicestershire LE11 3TU
1-year MSc in Environmental Management for Developing Countries

University of Luton, Park Square, Luton LU1 3JU
3-year BSc/BSc (Hons) in Environmental Management

Manchester Metropolitan University, All Saints, Manchester M13 9PL
2-year HND in Science (Environmental Analysis and Monitoring)
3-year BSc/BSc (Hons) in Environmental Management
1-year Foundation Environment Management
4-term MSc in Public and Environmental Health

University of Manchester, Oxford Road, Manchester M13 9PL
1-year MSc in Public Health and Epidemiology

University of Manchester Institute of Science and Technology (UMIST), PO Box 88, Manchester M60 1QD
3-year BEng (Hons) in Civil Engineering and Environmental Management
1 to 2-year MSc/DipTechSci in Pollution and Environmental Control

Middlesex University, White Hart Lane, London N17 8HR
3-year BA/BSc (Hons) Environmental Management and Policy
3-year BA (Hons) in Environment and Business Management

Nene College, Moulton Park, Northampton NN2 7AL
3-year BSc (Hons) in Environmental Science (Monitoring and Management)

Newbury College, Oxford Road, Newbury, Berkshire RE13 1PQ
1-year BSc/BSc (Hons) in Maritime Environmental Management

University of Newcastle upon Tyne, Newcastle upon Tyne NE1 7RU
1-year MSc/Diploma in Tropical Coastal Management

Newton Rigg College, Newton Rigg, Penrith, Cumbria CA11 0AH
2-year HND in Environmental Land Management

Normal College, Bangor, Gwynedd LL57 2PX
3-year BA/BA (Hons) in Environmental Planning and Management

North East Surrey College of Technology, Reigate Road, Ewell, Surrey KT17 3DS
2-year HND in Environment Monitoring and Control
3-year BSc (Hons) in Environmental Management

North East Wales Institute, Newi Plascoch, Mold Road, Wrexham, Clwyd LL11 2AW
3-year BSc/BSc (Hons) in Environmental Management

Nottingham Trent University, Burton Street, Nottingham NG1 4BU
3-year BSc (Hons) in Combined Studies in Sciences

University of Nottingham, University Park, Nottingham NG7 2RD
3-year BEng (Hons) in Environmental Engineering and Resource Management

Open University, PO Box 200, Walton Hall, Milton Keynes, Buckinghamshire MK7 6YZ
Distance learning
Environmental Control and Public Health
Environmental Monitoring and Control

Otley College of Agriculture and Horticulture, Otley, Ipswich, Suffolk IP6 9EY
2-year Advanced National Certificate in Conservation Management

Oxford Brookes University, Gipsy Lane, Headington, Oxford OX3 0BP
1-year MSc/Diploma in Environmental Assessment and Management

University of Oxford, University Offices, Wellington Square, Oxford OX1 2JD
1-year MSc in Environmental Change and Management

University of Paisley, Paisley Campus, High Street, Paisley PA1 2BE
1-year MSc/PgDip in Environmental Management

University of Portsmouth, Museum Road, Portsmouth PO1 2QQ
1-year MSc/PgDip/PgCert in Coastal and Marine Resource Management

Robert Gordon University, Schoolhill, Aberdeen AB9 1FR
1-year MSc/Diploma in Ecological Design

Salford University College, Frederick Road, Salford M6 6PU
2-year HND in Environmental Health

South Bank University, 103 Borough Road, London SE1 0AA
1-year Foundation in Occupational and Environmental Health and Safety
3-year BSc (Hons) in Environmental Biology
1-year MSc/PgDip in Environmental Monitoring and Assessment

Southampton Institute of Higher Education, East Park Terrace, Southampton SO9 4WW
4-year BSc (Hons) in Maritime Environmental Management

University of Stirling, Stirling FK9 4LA
1-year MSc/Diploma in Environmental Management
1-year MSc/PgDip in Environmental Health

University of Strathclyde, Glasgow G1 1XQ
4-year BSc/BSc (Hons) in Environmental Health
12 to 21-month MSc/PgDip in Environmental Health

University of Sunderland, Langham Tower, Ryhope Road, Sunderland SR2 7EE
2-year HND in Environmental Management
1-year MSc/Diploma in Environmental Management

University of Surrey, Guildford, Surrey GU2 5XH
1-year MSc/PgDip in Environmental Health: Water Surveillance and Quality Management
4-year MSc Environmental Management (Conservation Management, Pollution Control)

University College London, Gower Street, London WC1E 6BT
9-month MSc in Environmental Design and Engineering BSE.D

University College Stockton on Tees, University Boulevard, Thornaby, Stockton on Tees TS17 6BH
3-year BSc/BSc (Hons) in Environmental Management

University of Warwick, Coventry CV4 7AL

1-year MSc in Ecosystems, Analysis and Resource Management

University of the West of England, Frenchay Campus, Coldharbour Lane, Bristol BS16 1QY
3-year BSc (Hons) in Geography and Environmental Management
3-year BSc (Hons) in Environmental Quality and Resource Management
3-year BSc (Hons) in Environmental Health

Wye College, Wye, Ashford, Kent TN25 5AH
1-year MSc in Rural Resource and Environmental Policy

University of York, Heslington, York YO1 5DD
BSc (Hons) in Environmental Economics and Environmental Management
MSc/Diploma in Environmental Economics and Environmental Management

Landscape Architecture

University of Birmingham, Edgbaston, Birmingham B15 2TT
1-year MSc/Diploma in Landscape Development and Planning

University of Central England in Birmingham, Perry Barr, Birmingham B42 2SU
3-year BA/BA (Hons) Landscape Architecture
4-year MA/PgDip in Landscape Architecture

Cheltenham and Gloucester College of Higher Education, PO Box 220, The Park, Cheltenham, Glos GL50 4QF
3-year BA (Hons) in Landscape Architecture
1-year MA/Diploma/Certificate in Landscape Architecture
1-year MA/Diploma/Certificate in Landscape and Society

University of Edinburgh, Edinburgh EH8 9YL
2-year MLA in Landscape Architecture
1-year Diploma in Landscape Studies

University of Greenwich, Wellington Street, London SE18 6PF
3-year BA/BA (Hons) in Landscape Architecture
1-year Hons Diploma in Landscape Architecture
1-year MA in Landscape Architecture

Heriot-Watt University, Faculty of Environmental Studies, Lauriston Place, Edinburgh EH3 9DF
5-year BA (Hons) in Landscape Architecture
4-term MLA/Diploma in Landscape Architecture
4-term MSc/Diploma in Landscape Resources

Kingston University, Penrhyn Road, Kingston upon Thames, Surrey KT1 2EE
4-year BA (Hons) in Landscape Studies

Leeds Metropolitan University, Landscape Architecture, School of the Environment, Brunswick Building, Leeds LS2 8BU
3-year BA (Hons) in Landscape Design
1-year PgDip in Landscape Architecture

University of Leicester, University Road, Leicester LE1 7RH
1-year MA/PgDip in Landscape Studies

Manchester Metropolitan University, Manchester M15 6BH
3-year BA (Hons) in Landscape Design
1-year BLandArch in Landscape Architecture

University of Manchester, Oxford Road, Manchester M13 9PL
2-year MLD/BLD/Diploma in Landscape Design
1-year MA(LM)/DipLM in Landscape Management

University of Newcastle upon Tyne, Newcastle upon Tyne NE1 7RU
2-year MLD in Landscape Design
1-year MA in Landscape Design Studies

Oxford Brookes University, Gipsy Lane, Headington, Oxford OX3 0BP
1-year MA/Diploma in European Landscape Planning

University of Reading, PO Box 217, Reading, Berkshire RG6 2AH
4-year BSc (Hons) in Landscape Management

University of Sheffield, Department of Landscape Architecture, Sheffield S10 2TN
3-year Bsc (Hons) in Landscape Design and Archaeology
3-year BSc (Hons) in Landscape Design and Plant Science
1 to 2-year March/March Studies in Landscape
1 to 2-year MA/Diploma in Landscape Design
1-year Certificate in Landscape Studies

Wye College, Wye, Ashford, Kent TN25 5AH
1-year MSc in Landscape Ecology, Design and Management

Environmental Planning

University of Aberdeen, University Office, Regent Walk, Aberdeen AB9 1FX
1-year MSc/PgDip in Rural and Regional Resources Planning

Angla Polytechnic University, Victoria Road South, Chelmsford, Essex CM1 1LL
3-year BSc (Hons) in Planning and Development

University of Wales, College of Cardiff, Aberconway Building, Colum Drive, Cardiff CF1 3YN
3-year BSc (Hons) in City and Regional Planning
2-year MSc in City and Regional Planning
1-year MSc in Urban Planning in Developing Countries

University of Central England in Birmingham, Perry Barr, Birmingham B42 2SU
4-year BSc (Hons) and Diploma in Environmental Planning
1½-year MSc/PgDip in Town and Country Planning

Cheltenham and Gloucester College of Higher Education, PO Box 220, The Park, Cheltenham, Gloucestershire GL50 2QF
3-year BA (Hons) in Countryside Planning

Coventry University, Priory Street, Coventry CV1 5FB
3-year BSc (Hons) in Built Environment Studies
1-year MA/Diploma in Regional Planning

Cranfield University, Silsoe College, Silsoe, Bedfordshire MK45 4DT
9-month PgDip in Land Resource Planning

Croydon College, Fairfield, Croydon, Surrey CR9 1DX
1-year BA in Foundation Town Planning

Duncan of Jordanstone College of Art, 13 Perth Road, Dundee DD1 4HT
4-year BSc (Hons) in Town and Regional Planning
4-term MSc/Diploma in European Urban Conservation

University of Dundee, Department of Town and Regional Planning, Perth Road, Dundee DD1 4HN
4-year BSc (Hons) in Town and Regional Planning

University of Greenwich, Wellington Street, London SE18 6PF
1-year MA in Urban Design

Edinburgh College of Art, Heriot-Watt University, Lauriston Place, Edinburgh EH3 9DF
5-year BSc (Hons) in Town Planning
1-year MSc/Diploma in Urban Design
2-year MURP in Urban and Regional Planning
2-year PgDip in Town and Country Planning
9-month Diploma in Planning Studies (Developing Countries)

Liverpool John Moores University, Rodney House, 70 Mount Pleasant, Liverpool L3 5UX
12-month MSc in Urban Renewal (Regeneration and Design)

University of Liverpool, PO Box 147, Liverpool L69 3BX
1-year MA in Metropolitan Planning

London School of Economics, Houghton Street, London WC2A 2AE
3-year BSc (Econ) (Hons) in Environment and Planning

Loughborough University of Technology, Loughborough, Leicestershire LE11 3TU
3-year BSc (Hons) in Transport Management and Planning
1-year MSc in Regional and Urban Planning Studies

University of Manchester, Department of Planning and Landscape, Manchester M13 9PL
3-year BA (Hons) in Town and Country Planning
2-year MTPI/BTP/Diploma in Town and Country Planning

Mid Kent College of Higher and Further Education, Horsted, Maidstone Road, Chatham, Kent ME5 9UQ
1-year BA/BA (Hons) in Town Planning (Foundation year)

University of Newcastle upon Tyne, Department of Town and Country Planning, Claremont Tower, Claremont Road, Newcastle upon Tyne NE1 7RU
5-year BA (Hons)/Diploma in Town Planning
1-year MA/Diploma in Urban Design
1-year MA in Planning Studies
2-year MTP in Town Planning
12-month Diploma in Town and Country Planning

The Normal College, Bangor, Gwynedd LL57 2PX
3-year BA/BA (Hons) in Environmental Planning and Management

North East Wales Institute of Higher Education, Newi Plascoch, Mold Road, Wrexham, Clwyd LL11 2AW
BSc/BSc (Hons) in Planning and Development

University of Nottingham, University Park, Nottingham NG7 2RD
3-year BA (Hons) in Urban Planning and Management
21-month MA/Diploma in Environmental Planning for Developing Countries
2-year MA/Diploma in Environmental Planning
9-month MA in Urban Design (for Architects)

Oxford Brookes University, School of Planning, Headington, Oxford OX3 0BP
4-year BA (Hons) with Diploma in Planning Studies
2-year MSc/Diploma in Urban Planning
1-year MSc in Planning Studies
1-year Diploma in Planning
1-year Diploma in Urban Planning Studies

The Queen's University of Belfast, School of Architecture and Planning, 2 Lennoxvale, Belfast BT9 5BY
3-year BSc in Environmental Planning
1-year MSc/Diploma in Town and Country Planning

University of Reading, Whiteknights, PO Box 217, Reading, Berkshire RG5 2AH
2-year MPhil/Advanced Diploma in Environmental Planning

Sheffield Hallam University, Pond Street, Sheffield S1 1WB
3-year BA (Hons) in Planning Studies/Town Planning

University of Sheffield, PO Box 594, Sheffield S10 2UH
2-year MA/Diploma in Town and Regional Planning
9-month Diploma in Planning Studies

South Bank University, Department of Planning, Housing and Development, Wandsworth Road, London SW8 2JZ
3-year BA (Hons) in Town Planning Studies
1-year Foundation Town Planning Studies
2-year MA/PgDip in Town Planning

University of Strathclyde, Jordanhill Campus, Southbrae Drive, Glasgow G13 1PP
3-year BA/BA (Hons) in Geography and Planning
4-year BA (Hons) in Planning
24-month MSc/PgDip in Urban and Regional Planning
12-month MSc/PgDip in Urban Design

University College London, Gower Street, London WC1E 6BT
3-year BSc (Hons) in Town and Country Planning

University of Wales, Singleton Park, Swansea SA2 8PP
1-year MSc in Urban Development Planning
1-year MSc in Environmental Design and Engineering
1-year Diploma in Urban Planning Practice for Developing Countries
12-month MSc in Regional Development Planning

University of the West of England, Frenchay Campus, Coldharbour Lane, Bristol BS16 1QY
3-year BA (Hons) in Town and Country Planning
2-year MA/PgDip in Town and Country Planning

University of Westminster, School of Urban Development and Planning, 35 Marylebone Road, London NW1 5LS
1-year Foundation Town Planning
3-year BA/BA (Hons) in Town Planning
1-year Foundation European Planning with a Language
3-year BA/BA (Hons) in Tourism and Planning
1-year PgDip in Urban Planning

Chapter 13

Useful Reading

There are numerous journals that deal with conservation and the natural environment. The most important are listed in this chapter. Some journals carry job advertisements, but they are included here principally because they give in-depth coverage of their topics and are useful sources of news and information.

Journals

Conservation and Natural History

BBC Wildlife Magazine
World Publications, Worldlands, 80 Wood Lane, London W12 0TT; 0181 743 5588. Monthly.

Birds
Royal Society for the Protection of Birds, The Lodge, Sandy, Bedfordshire SG12 2DL; 01767 680551. Quarterly.

The Conserver
British Trust for Conservation Volunteers, 36 St Mary's Street, Wallingford, Oxfordshire OX10 0EU; 01491 39766. Quarterly.

Country-Side
British Naturalists Association, 30 Lawson Street, Kettering, Northamptonshire NN16 8XH; 01539 411636. Three issues per year.

Green Drum
18 Cofton Lake Road, Birmingham B45 8PL; 0121 445 2576. Quarterly.

Green Magazine
The Northern and Shell Building, PO Box 381, Millharbour, London E14 9TW; 0171 987 5090.

The National Trust
National Trust, 36 Queen Anne's Gate, London SW1H 0AS; 0171 222 9251. Three issues per year.

Natural World
The Illustrated London News, Sea Containers House, 20 Upper Ground, London SE1 9PF; 0171 928 2111. Three issues per year.

Oryx
Blackwell Scientific Publications, Osney Mead, Oxford OX2 0EL; 01865 240201. Quarterly.

Trends in Ecology and Evolution
Elsevier Trends Journals, 68 Hills Road, Cambridge CB2 1LA; 01223 315961. Monthly.

Wildfowl & Wetlands
The Wildfowl and Wetlands Trust, Slimbridge, Gloucestershire GL2 7BT; 01453 890333. Two issues per year.

WWF News
World Wide Fund for Nature, Panda House, Weyside Park, Godalming, Surrey GU7 1XR; 01483 426444. Quarterly.

Environment

The Ecologist
Agriculture House, Bath Road, Sturminster Newton, Dorset DT10 1DU; 01258 73476. Bi-monthly.

Green World
Green Party, 49 York Road, Aldershot, Hampshire GU11 3JQ; 01252 330506. Six issues per year.

Journal of Ecology
Blackwell Scientific Publications Ltd, Osney Mead, Oxford OX2 0EL; 01865 240201. Quarterly.

Resurgence
Resurgence Trust, Ford House, Hartland, Bideford, Devon EX39 6EE; 01237 441293. Bi-monthly.

Resource Management

Commonwealth Forestry Handbook
Commonwealth Forestry Association, South Parks Road, Oxford OX1 3RB; 01865 275072. Quarterly.

Forestry
Oxford University Press, Journals Department, Walton Street, Oxford OX2 6DP; 01865 56767. Four issues per year.

Landscape Design
13a West Street, Reigate, Surrey RH2 9BL; 01737 225374. Monthly.

The Planner
26 Portland Place, London W1N 4BE; 0171 636 9107. Monthly.

Quarterly Journal of Forestry
Royal Forestry Society of England, Wales and Northern Ireland, 102 High Street, Tring, Hertfordshire HP23 4AF; 01442 822028. Quarterly.

Resources Policy
Butterworth-Heinemann Companies, Linacre House, Jordan Hill, Oxford OX2 8DP; 01865 310366.

Scottish Forestry
Royal Scottish Forestry Society, Camsie House, Charleston, Dunfermline, Fife KY11 3EE. 01383 873014. Quarterly.

Town & Country Planning
Town & Country Planning Association, 17 Carlton House Terrace, London SW1Y 5AS; 0171 930 8903. Monthly.

World Water & Environmental Engineer
Thomas Telford Ltd, Thomas Telford House, 1 Heron Quay, London E14 9XF; 0171 987 6999. Monthly.

Science

The Biochemical Journal
Portland Press, 59 Portland Place, London W1N 3AJ; 0171 580 5530. Bi-monthly.

Biological Conservation
Elsevier Applied Science, Crown House, Linton Road, Barking, Essex IG11 8JU; 0181 594 7272. Monthly.

Biologist
Institute of Biology, 20 Queensberry Place, London SW7 2DZ; 0171 581 8333. Five issues per year.

Journal of Animal Ecology
Blackwell Scientific Publications Ltd, Osney Mead, Oxford OX2 0EL; 01865 240201. Three issues per year.

Journal of Applied Ecology
Blackwell Scientific Publications Ltd, Osney Mead, Oxford OX2 0EL; 01865 240201. Quarterly.

Journal of Biological Education
Institute of Biology, 20 Queensberry Place, London SW7 2DZ; 0171 581 8333. Quarterly.

Journal of Ecology
Blackwell Scientific Publications Ltd, Osney Mead, Oxford OX2 0EL; 01865 240201. Quarterly.

Journal of Natural History
Taylor & Francis Ltd, 4 John Street, London WC1N 2ET; 0171 405 2237. Bi-monthly.

Journal of the Marine Biological Association
Cambridge University Press, The Edinburgh Building, Shaftesbury Road, Cambridge CB2 2RU; 01223 312393. Quarterly.

Journal of Zoology
Oxford University Press, Walton Street, Oxford OX28 6DP; 01865 882890. Monthly.

Mammal Review
Blackwell Scientific Publications Ltd, Osney Mead, Oxford OX2 0EL; 01865 240201. Quarterly.

Marine Environmental Research
Elsevier Applied Science, Crown House, Linton Road, Barking, Essex IG11 8JU; 0181 594 7272. Eight issues per year.

Marine Pollution Bulletin
Pergamon Press plc, Headington Hill Hall, Oxford OX3 0BW; 01865 794141. Fortnightly.

Nature
Macmillan Magazines, 4 Little Essex Street, London WC2R 3LF; 0171 836 6633. Weekly.

New Scientist
IPC Magazines, King's Reach Tower, Stamford Street, London SE1 9LS; 071-261 5000. Weekly.

Conservation of the Built Environment

Urban Focus
Civic Trust, 17 Carlton House Terrace, London SW1Y 5AW;
0171 930 0914. Quarterly.

Global Issues

New Internationalist
New Internationalist Publications Ltd, 55 Rectory Road, Oxford OX4 1BW; 01865 728181. Monthly.

Third World Planning Review
Liverpool University Press, PO Box 147, Liverpool L69 3BX; 0151 794 2237. Quarterly.

Third World Quarterly
Carfax Publishing Co, PO Box 25, Abingdon, Oxon OX13 6BS; 01235 555335. Quarterly.

World Development
Pergamon Press plc, Headington Hill Hall, Oxford OX3 0BW; 01865 794141. Monthly.

Foreign Journals

Environmental Management
Springer-Verlag New York Inc, 175 Fifth Avenue, New York, NY 10010, USA. Bi-monthly.

Sierra
Sierra Club, 730 Polk Street, San Francisco, California 94109, USA.

Smithsonian
Arts and Industries Building, 900 Jefferson Drive, Washington DC 20560, USA.

Reference and General Books

Careers in Conservation, RSPB Youth Unit, The Lodge, Sandy, Beds SG19 2DL

Caring for the Earth, WWF UK in association with Earthscan Publications (1991)

Countryside Education and Training Directory, Countryside Commission, 19–23 Albert Road, Manchester M19 2EQ

Directory of Environmental Courses, The Environment Council, 80 York Way, London N1 9AG

The Earth Summit Agreements, Earthscan Publications (1993)

Environmental Careers Handbook, Institution of Environmental Science, Trotman (1993)

Environmental Responsibility: An agenda for further and higher education, Peter Toyne, HMSO (1993)

Green Business, Malcolm Wheatley, WWF UK and Pitman Publishing with the Institute of Management

Greening the Curriculum, WWF UK (1991)

Media Courses UK, British Film Institute, 21 Stephen Street, London W1P 1PL

Who's Who in the Environment, The Environment Council (1993)

Chapter 14

Useful Addresses

Listed in this chapter are the addresses of many organisations referred to in Chapters 1 to 10, together with those of other relevant and related bodies. Some are potential employers, some accept members and voluntary assistance, some are professional institutes; all are useful sources of further information on different aspects of conservation. When writing to the voluntary bodies, *enclose a stamped addressed envelope.*

Voluntary and Non-Governmental Bodies

British Association of Landscape Industries
9 Henry Street, Keighley, West Yorkshire BD21 3DR; 01535 606139

British Association of Nature Conservationists
PO Box 14, Neston, South Wirral, Merseyside L94 7UP

British Trust for Conservation Volunteers
36 St Mary's Street, Wallingford, Oxfordshire OX10 0EU; 01491 39766

British Trust for Ornithology
The Nunnery, Nunnery Place, Thetford, Norfolk IP24 2PU; 01842 750050

Centre for Alternative Technology
Machynlleth, Powys SY20 9AZ; 01654 702400

Civic Trust
17 Carlton House Terrace, London SW1Y 5AW; 0171 930 0914

Conservation Trust
National Environment Education Centre, George Palmer Site, Northumberland Avenue, Reading RG2 7PW; 01734 868442

Council for National Parks
246 Lavender Hill, London SW11 1LJ; 0171 924 4044

Council for the Protection of Rural England
Warwick House, 25 Buckingham Palace Road, London SW1W 0PP; 0171 976 6433

English Heritage
23 Savile Row, London W1X 2HE; 0171 973 3000

The Environment Council
21 Elizabeth Street, London SW1W 9RP; 0171 824 8411

Environmental Investigation Agency
2 Pear Tree Court, London EC1R 0DS; 0171 490 7040

Friends of the Earth
26–28 Underwood Street, London N1 7JQ; 0171 490 1555

Greenpeace Environmental Trust
Canonbury Villas, London N1 2PN; 0171 354 5100

Henry Doubleday Research Association
Ryton Gardens, Ryton-on-Dunsmore, Coventry CB8 3LG; 01203 303517

Marine Conservation Society
9 Gloucester Road, Ross-on-Wye, Herefordshire HR9 5BU; 01989 566017

National Society for Clean Air
136 North Street, Brighton, Sussex BN1 1RG; 01273 26313

National Trust
36 Queen Anne's Gate, London SW1H 9AS; 0171 222 9251

National Trust for Ireland
Tailor's Hall, Back Lane, Dublin 2, Republic of Ireland; 353 (1) 541786

National Trust for Scotland
5 Charlotte Square, Edinburgh EH2 4DU; 0131 226 5922

Organic Growers Association
86 Colston Street, Bristol BS1 5BB; 0117 9299800

Population Concern
231 Tottenham Court Road, London WC1P 9AE; 0171 387 0455

Royal Society for the Protection of Birds
The Lodge, Sandy, Bedfordshire SG19 2DL; 01767 680551

RSNC The Wildlife Trusts Partnership
The Green, Wickham Park, Lincoln LN5 5RR; 01522 544400

Scottish Civic Trust
24 George Square, Glasgow G2 1EF; 0141 221 1466

Scottish Wildlife Trust Ltd
Cramond House, Kirk Cramond, Cramond Glebe Road, Edinburgh EH4 6NS; 0131 312 7765

Soil Association Ltd
86 Colston Street, Bristol BS1 5BB; 0117 9290661

The Tidy Britain Group
The Pier, Wigan WN3 4EX; 01942 824620

Tree Council
35 Belgrave Square, London SW1X 8QN; 0171 235 8854

Whale & Dolphin Conservation Society
19a James Street West, Bath, Avon BA1 2BT; 01225 334511

Wild Flower Society
68 Outwood Road, Loughborough, Leicestershire LE11 3LY; 01509 215598

Wildfowl and Wetlands Trust
The New Grounds, Slimbridge, Gloucestershire GL2 7BT; 01453 890333

Woodland Trust
Autumn Park, Dysart Road, Grantham, Lincolnshire NG31 6LL; 01476 74297

WWF UK (World Wide Fund For Nature)
Panda House, Weyside Park, Godalming, Surrey GU7 1BR; 0148342 6444

Resource Management

British Wind Energy Association
42 Kingsway, London WC2B 6EX; 0171 404 3433

Institute of Chartered Foresters
7A Colme Street, Edinburgh EH3 6AA; 0131 225 2705

Landscape Institute
6-7 Barnard Mews, London SW11 1QU; 0171 738 9166

Network for Alternative Technology and Technology Assessment (NATTA)
c/o Energy and Environment Research Unit, Faculty of Technology, The Open University, Walton Hall, Milton Keynes MK7 6AA

Royal Forestry Society of England and Wales
102 High Street, Tring, Hertfordshire HP23 4AF; 01442 822028

Royal Institution of Chartered Surveyors
Rural Practice Division, 12 Great George Street,
Parliament Square, London SW1P 3AD; 0171 222 7000

Royal Scottish Forestry Society
Camsie House, Charlestown, Dunfermline, Fife KY11 3EE; 01383 873014

Royal Town Planning Institute
26 Portland Place, London W1N 4BE; 0171 636 9107

Timber Growers United Kingdom Ltd
Agriculture House, Knightsbridge, London SW1X 7NJ; 0171 235 2925

Water Authorities Association
1 Queen Anne's Gate, London SW1H 9BT; 0171 222 8111

Government Bodies

Agricultural Development and Advisory Service
MAFF, Victory House, 30-34 Kingsway, London WC2B 6TU; 0171 405 4310

Agricultural Scientific Services Station
East Craigs, Corstorphine, Edinburgh EH12 8NJ; 0131 556 8400

British Antarctic Survey
High Cross, Madingley Road, Cambridge CB3 0ET; 01223 311354

British Geological Survey
London Information Desk
Geological Museum, Exhibition Road, London SW7; 0171 589 4090

Department for Education
Sanctuary Buildings, Great Smith Street, London SW1P 3BT; 0171 925 5000

Department of the Environment
2 Marsham Street, London SW1P 3EB; 0171 276 3000

Energy Efficiency Office
1 Palace Street, London SW1E 5HE; 0171 238 3000

Farming and Wildlife Advisory Group
National Agricultural Centre, Stoneleigh, Kenilworth CV8 2RX; 01203 696699

Forestry Commission
231 Corstorphine Road, Edinburgh EH12 7AT; 0131 334 0303

Forestry and Arboricultural Safety and Training Council
231 Corstorphine Road, Edinburgh EH12 7AT; 0131 334 8083

Natural Environment Research Council
Polaris House, North StarAvenue, Swindon SW2 1EU; 01793 411500

Recruitment and Assessment Services, Civil Service
Alençon Link, Basingstoke, Hampshire RG21 1JB; 01256 29222

Statutory Bodies

Countryside Commission
John Dower House, Crescent Place, Cheltenham, Gloucestershire GL50 3RA; 01242 521381

Scottish Natural Heritage (Training)
Battleby, Redgorton, Perth PH1 3EW; 01738 27921

English Nature
Northminster House, Peterborough PE1 1UA; 01733 340345

Education and Research

Council for Environmental Education
University of Reading, London Road, Reading, Berkshire RG1 5AQ; 01734 756061

Countryside Council for Wales
Plas Penrhos, Ffordd Penrhos, Bangor, Gwynedd LL57 2LQ; 01248 370444

Dartington Hall, Totnes, Devon TQ9 6EL; 01803 862271

Field Studies Council
Preston Montford, Montford Bridge, Shrewsbury, Shropshire SY4 1HW; 01743 850164

National Rivers Authority
30-34 Albert Embankment, London SE1 4TL; 0171 820 0101

Scottish Field Studies Association Ltd
Kindrogan Field Centre, Enochdhn, Blairgowrie, Perthshire PH10 7PG; 01250 881286

Scottish Natural Heritage (Recruitment)
12 Hope Terrace, Edinburgh EH9 2AS; 0131 447 4784

UCAS (Universities & Colleges Admissions Service)
Fulton House, Jessop Avenue, Cheltenham, Gloucestershire GL50 3SH; 01242 222444

Science Bodies

British Ecological Society
Burlington House, Piccadilly, London W1V 0LQ; 0171 434 2641

Institute of Biology
20 Queensberry Place, London SW7 2DZ; 0171 581 8333

Institution of Environmental Sciences
14 Prince's Gate, London SW7 1PU; 0171 766 6755

Royal Horticultural Society
Vincent Square, London SW1P 2PE; 0171 834 4333. Hon Secretary: 01252 515511

Zoological Society of London
Zoological Gardens, Regent's Park, London NW1 4RY; 0171 722 3333

International

Catholic Institute for International Relations
Unit 3 Canonbury Yard, 190A New North Road, London N1 7BJ; 0171 354 0883

Centre for World Development Education
1 Catton Street, London WC1R 4AB; 0171 831 3844

Commonwealth Secretariat
Marlborough House, London SW1Y 5HX; 0171 839 3411

Earthscan
120 Pentonville Road, London N1 9JN; 0171 278 0433

International Farm Experience Programme
Young Farmers' Clubs Centre, National Agricultural Centre, Stoneleigh, Kenilworth CV8 2LG; 01203 696584

International Institute for Environment and Development
3 Endsleigh Street, London WC1H 0DD; 0171 388 2177

International Planned Parenthood Federation
Regents College, Regent's Park, London NW1 4NS; 0171 486 0741

National Council for Voluntary Organisations
Regents Wharf, All Saints Road, London N1 9RL; 0171 713 6161

Overseas Development Administration
Esso House, Victoria Street, London SW1E 5JW; 0171 917 7000

Overseas Development Natural Resources Institute
56 Grays Inn Road, London WC1X 8LT; 0171 405 7943

Panos Institute
9 White Lion Street, London N1 9PD; 0171 278 1111

Population Concern
231 Tottenham Court Road, London W1P 9AE; 0171 631 1546

United Nations Association of Great Britain & Northern Ireland
3 Whitehall Court, London SW1A 2EL; 0171 930 2931

United Nations Information Centre
Ship House, 20 Buckingham Gate, London SW1E 6LB; 0171 630 1981

Voluntary Service Overseas
317 Putney Bridge Road, London SW15 2PN; 0181 780 2266